HARNESS: TYPES AND USAGE FOR RIDING

- DRIVING AND CARRIAGE HORSES -

WITH REMARKS ON CRACTION, AND

THE USE OF THE CAPE CART

BY

JOHN PHILIPSON,

ALSO, AN APPENDIX BY THE SAME AUTHOR,
CONTAINING SOME VERY
IMPORTANT DIRECTIONS TO GROOMS AND
COACHMEN RESPECTING THEIR DUTIES, THEIR
DRESS, AND HINTS ON DRIVING.

British Library Cataloguing-in-Publication Data
A catalogue record for this book is available from the
British Library

Contents

Horses – Sports and Utility

The horse (Equus ferus caballus) is one of two extant subspecies of Equus ferus. It is an odd-toed ungulate mammal belonging to the taxonomic family 'Equidae'. The horse has evolved over the past 45 to 55 million years from a small multi-toed creature into the large, single-toed animal of today. Humans began to domesticate horses around 4000 BC, and their domestication is believed to have been widespread by 3000 BC. We, as humans have interacted with horses in a multitude of ways throughout history – from sport competitions and non-competitive recreational pursuits, to working activities such as police work, agriculture, entertainment and therapy. Horses have also been used in warfare, from which a wide variety of riding and driving techniques developed, using many different styles of equipment and methods of control. With this range of uses in mind, there is an equally extensive, specialized vocabulary used to describe equine-related concepts, covering everything from anatomy to life stages, size, colours, markings, breeds, locomotion, and behaviour.

Sporting events are some of the largest and best-known activities involving horses, and here – communication between human and horse is paramount. To aid this process, horses are usually ridden with a saddle on their backs to assist the rider with balance and positioning, and a bridle or related headgear to assist the rider in maintaining control. Historically, equestrians honed their craft through games and races; providing skills needed for battle, as well as entertainment for home crowds. Today, these competitions have evolved into racing, dressage, eventing and show jumping – many of which have their origins in military training, focused on control and balance of both the horse and rider. Other sports, such as rodeo, developed from practical skills such as those needed on working ranches and stations. Horse racing of all types evolved from impromptu competitions between riders or drivers, and has since become a multi-million pound industry. It is watched in almost every nation of the world, in its three main forms: 'flat racing' (long, even stretches), 'steeplechasing' (racing over jumps) and 'harness racing' (where horses trot or pace whilst pulling a driver in a small, light cart). A major part of horse racing's economic importance lies in the gambling associated with it.

All forms of competition, requiring demanding and specialized skills from both horse and rider, resulted in the systematic development of specialized breeds and equipment for each sport. Horse shows, which have their origins in medieval European fairs, are held around the world. They host a huge range of classes, covering all of the mounted and harness disciplines, as well as 'In-hand' classes where the horses are led, rather than ridden, to be evaluated on their conformation. The method of judging varies with the discipline, but winning usually depends on style and ability of both horse and rider. Sports such as polo do not judge the horse itself, but rather use the horse as a partner for human competitors as a necessary part of the game. Although the horse requires specialized training to participate, the details of its performance are not judged, only the result of the rider's actions—be it getting a ball through a goal or some other task. A similar, historical example of sports partnerships between human and horse is 'jousting', in which the main goal is for one rider to unseat the other. This pastime is still practiced by some sportsmen today.

There are certain jobs that horses do very well, and no technology has yet developed to fully replace them. For example, mounted police horses are still effective for certain types of patrol duties and crowd control. Cattle ranches still require riders on horseback to round up cattle that are scattered across remote, rugged terrain. In more urban areas, horses used to be the main form of transport, in the form of pulling carriages, and are still extensively used (especially in the UK) for ceremonial functions, i.e. horse-drawn carriages transporting dignitaries, military personnel or even the royal family. Horses can also be used in areas where it is necessary to avoid vehicular disruption to delicate soil, such as nature reserves. They may also be the only form of transport allowed in wilderness areas, often because of the fact that horses are quieter than motorised vehicles, therefore impacting less on their surroundings. Although machinery has replaced horses in many parts of the world, an estimated 100 million horses, donkeys and mules are still used for agriculture and transportation in less developed areas. This number includes around 27 million working animals in Africa alone.

As well as these labour intensive uses, horses can also be incredibly valuable for therapy. People of all ages with physical and mental disabilities obtain beneficial results from association with horses. Therapeutic riding is used to mentally and physically stimulate disabled persons and help them improve their lives through improved balance and coordination, increased self-confidence, and a greater feeling of freedom and independence. Horses also provide psychological benefits to people whether they actually ride or not. 'Equine-assisted' or 'equine-facilitated' therapy is a form of experiential psychotherapy that uses horses as companion animals to assist people with mental illness, including anxiety disorders, psychotic disorders, mood disorders, behavioural difficulties, and those who are going through major life changes. There are also experimental programs using horses in prison settings. Exposure to horses appears to improve the behaviour of inmates and help reduce recidivism when they leave.

As a concluding note, one of the most important aspects of equine care is farriery; a specialist in equine hoof care. Horses aid humans in so many ways, it is important to ensure that they are properly equipped and cared for. Farriers have largely replaced blacksmiths (after this specialism mostly became redundant after the industrial revolution), and are highly skilled in both metalwork and horse anatomy. Historically, the jobs of farrier and blacksmith were practically synonymous, shown by the etymology of the word: farrier comes from Middle French ferrier (blacksmith), and from the Latin word ferrum (iron). Modern day farriers usually specialize in horseshoeing though, focusing their time and effort on the care of the horse's hoof, including trimming and balancing of the hoof, as well as the placing of the shoes. Additional tasks for the farrier include dealing with injured or diseased hooves and application of special shoes for racing, training or 'cosmetic' purposes. In countries such as the United Kingdom, it is illegal for people other than registered farriers to call themselves a farrier or to carry out any farriery work, the primary aim being 'to prevent and avoid suffering by and cruelty to horses arising from the shoeing of horses by unskilled persons.' This is not the case in all countries however, where horse protection is severely lacking.

We hope the reader enjoys this book.

PREFACE.

ALTHOUGH it may be a truism to assert that the comfort of the horse should be the first consideration with us, it is very necessary for the writer to insist on the fact that, up to the present time, such a truism has not had sufficient practical acceptation. It is a disgrace to our boasted civilization that Mr. Smiles should ever have been able to write: "There is no slavery in England! But look at the 'bus and cab and cart horses, and you will find that slavery exists for horses. It was said by James Howell, Clerk of the Council, as long ago as 1642, that England is called 'The hell of horses, and not without cause.' Cabs are driven by worn-out animals, and one or more of their feet are full of pain."

Again he writes, "Ask the carriage horse, galled with its detestable bearing-rein, drawing the proud beauty along the Row, its mouth covered with foam and sometimes with blood; and what would it say? That men and women were alike its merciless tyrants. And yet such ladies go to anti-vivisection meetings to protest against cruelty to animals Fashion is strong—stronger I fear than humanity—but still I have hopes. Fashion no longer orders horses to be cropped, docked, and nicked; therefore these new forms of distortion and cruelty may give way. If a few ladies of fashion would join with men and women of common sense and lovers of humanity, we should soon wipe out this blot upon our civilization."

Mr. Flower, again, has contributed much towards a more humane practice of the use of the bearing-rein, but, alas! has not yet succeeded in putting down the gag.

Therefore, the writer would beg the reader to bear in mind that humanity, before everything in the use of harness, is the guiding principle on which the following little treatise has been written; whilst he has endeavoured to keep to the practical and the useful where certain details do not absolutely interfere with the horses' comfort.

In these days of education and technical colleges apprentices and others should be instructed in the uses of different parts of harness, the structure of the horse, &c.; and should these remarks tend, in however slight a degree, to an improved method of harnessing and greater comfort to the noblest friend of man, the writer will be amply repaid. Steps in the right direction are certainly being taken in different parts of the country in the establishment of annual processions of horses, in which valuable prizes are awarded to the *drivers* for careful grooming and harnessing, and the writer can speak with certainty as to the result of the May-day horse procession held in Newcastle. It would be difficult to imagine a more enjoyable sight than that presented on the first of May, 1882, when several thousand horses in splendid condition paraded the streets of the City, nor could the fact be ignored that the horses of the North-Eastern Railway Company (numbering about 300) were the pick of the exhibition, and these horses were worked without bearing-reins, winkers, or crupper docks. In presenting the prizes the Baroness Burdett Coutts gave expression to the sincere pleasure she had derived from the sight.

I have selected, as a frontispiece, "Ulysses ploughing the seashore" as beautifully representing the moral which I wish to pervade this essay, viz., that by kindness and good training a dumb animal can be brought under wonderful control, that simplicity in harness is desirable, and that the harness should be specially adapted for its work.

I am indebted to my daughters for many of the illustrations, none of which have any pretension to refined art, but are simply intended to exhibit particular features in my essay.

I take this opportunity to gratefully acknowledge many favours at the hands of several well-known North country gentlemen, and although I cannot enumerate all, it is only my duty to thank Mr. Jacob Wilson, of Chillingham; Mr. Richard Cail, Ex-Mayor of Newcastle; Mr. James Allport, of Derby; Mr. Inman, of the North-Eastern Railway; Mr. Marsters, of the East Coast Lines; and Captain Barthorp, of the London and North-Western Railway; all of whom have honoured me with most useful information, all the more valuable as coming from gentlemen whose exceptionally wide experience enables them to speak with the greatest authority.

For the rare woodcuts, by Thomas and John Bewick, I am indebted to the kindness of my friend Mr. Robert Robinson, bookseller.

JOHN PHILIPSON.

NEWCASTLE-UPON-TYNE,

October 10th, 1882.

ST. NICHOLAS CHURCH STEEPLE.

ULYSSES PLOUGHING THE SEA SHORE.

TO THE RIGHT HONOURABLE

THE EARL OF TANKERVILLE, P.C., D.L., &c., &c.

MY LORD,

The permission which your Lordship has given me to dedicate this work to you confers an honour and a pleasure, and permits also the performance of a duty. It is well known that your Lordship has ever evinced a keen interest in the training and general management of "The Horse," especially with a view to the kind and gentle treatment of that noble animal.

To no one could these pages, which may tend toward this high object, be more fitly inscribed by

Your Lordship's
Obedient and Humble Servant,
JOHN PHILIPSON.

CHAPTER I.

THE first consideration in approaching this subject must necessarily be the horse, in which the power and intelligence of the brute creation appears to have its highest development.

In the domestic state the horse is generous, docile, spirited, and yet obedient; adapted to the various purposes of pleasure and convenience, equally serviceable for draught, the field, or the course.

In 1792 my townsman, Thomas Bewick, the famous naturalist and reviver of wood engraving, said—

The various excellencies of this noble animal, the grandcur of his stature, the elegance and proportion of his parts, the beautiful smoothness of his skin, the variety and gracefulness of his motions, and, above all, his utility, entitle him to a precedence in the history of the brute creation.

Sidney Smith's horses came running to him when he entered the field, and his favourite young donkey, when he saw him or any member of his family, would set off with ears down and tail erect in full bray to meet them. Sir Walter Scott narrates that on visiting his daughter, Mrs. Lockhart, the two donkeys would run upon the approach of the laird and be stroked by Sir Walter.

The horses employed by our railway companies and contractors, and the Manchester trolly horses, evince an amount of intelligence and plodding determination truly wonderful and unequalled by any other animal, and we have only to watch their labours to be pro-foundly impressed by the facility with which they shunt railway carriages, draw loaded trucks and waggons, which, though running easily when once started, require immense power to set them in motion, and the judgment they display, in stepping aside from between the rails to be out of danger while their chains are being detached, is remarkable. The following is an extract from *Smiles' Life of George and Robert Stephenson:—*

Stockton and Darlington Railway opened September 27th, 1825.

The three Stephenson locomotive engines were from the first regularly employed to work the coal trains. . . . For some years, however, the principal haulage of the line was performed by horses. The inclination of the gradient being towards the sea, this was perhaps the cheapest mode of traction so long as the traffic was not very large. The horse drew the train along the level road until on reaching a descending gradient, down which the train ran by its own gravity, the animal was unharnessed, when wheeling round to the other end of the waggons, to which a "dandy cart" was attached, its bottom being only a few inches from the rail, and bringing its step into unison with the speed of the train, he lept nimbly into his place in the hind car, which was suitably fitted with a well-filled hay rack.

As an example of the marvellous power of memory with which they are endowed, I may repeat, in his own words, an incident related to me by the veteran Master of the North Durham Foxhounds, H. L. Maynard, Esq.:—

Some years ago I purchased a brown mare by "MacOrville," very highly bred and good looking, that had a fearfully bad character from being wicked and unmanageable, but I liked the look of the mare so much that I was determined to have her. She had thrown her riders and was to all intents spoilt when she came into my stable. I got her used to my voice and kind words, and, by degrees, walked her out, always kindly treating her, and after a time mounted her. She gradually became quiet, and before the hunting season was over my wife rode her, using a snaffle bridle. I kept her four seasons, and she proved to be one of the best hunters in Yorkshire. I sold her to the late John Duncombe Shafto for a long price. The mare had been in his possession two years before I saw her again, when my wife and myself met the Duke's hounds near Raby. About fifty yards from where we were standing Mr. Shafto appeared on the mare, when all at once, Mr. Shafto said, the mare threw up her head, took the bit in her mouth, and rushed up to my wife. She had heard her voice, and she rubbed her nose in the most affectionate way against her habit, and for the whole of the day Mr. Shafto could not keep the mare away from us.

Another incident, in which Mr. Maynard acted a similar part, will bear repetition:—

Some years ago Lord de Lisle had a very fine grey mare which had quite got the better of his man and had become very wicked. He wished me to buy her, for, he said, my man is so much afraid of the mare he dare not go up to feed her. I went over to his place and bought the mare for £50, and she came to my stable the most *frightened* animal I ever saw. By good and gentle treatment I soon had her perfectly quiet and rode her in a snaffle, and, after hunting her four seasons, I sold her to Lord Henry Bentinck for 300 guineas, and she gave perfect satisfaction.

As an instance of what a horse may be made to do by a resolute rider, I may mention that in July, 1839, during the Chartist riots in Newcastle-on-Tyne—Mr. John Fife was mayor of Newcastle at the time, and he was afterwards knighted for the judicious manner in which he quelled the riots—a Dragoon being determined to clear the Royal Arcade of the mob, passed through on horseback and rushed down the long flight of steps* which leads from it into the Manors.

Although it is necessary that I should refer to the structure of the horse, I shall not trouble the reader with a description of that animal's anatomy, but shall limit myself to a few words on those parts more especially connected with draught and harness. Although it may be considered an unnecessary precaution by some, I have given the names commonly applied to various parts of the horse on Fig. 1, Plate I., to assist the uninitiated in identifying such parts when reference is made to them. Fig. 1 represents a once famous draught horse, and is taken from a plate presented to the members of the Northumberland Agricultural Society by my relative, Joseph Snowball, Esq., of Seaton Burn House, Northumberland, as a model of a well-shaped draught horse. The eye has been shown by Mayhew to display special provisions and to closely resemble that of the camel, proving, if any proof were necessary, that he had his origin in tropical regions, and though the horse may not possess the power of seeing in the dark with the same facility as the cat, it is certain that his powers of vision in the gloom are much superior to those of man. Besides, the horse possesses the power of telescopic vision, a faculty which must be accepted as supporting this argument, for, as Mayhew says, this power could seldom be needed in these northern climes, but would be a needed protection on the far stretching level of the desert. By a protection of seven muscles the horse is enabled to turn his eye in any direction and to draw it out of reach of danger at will, and if by chance any foreign substance gets into the eye, these muscles are stimulated to sudden activity, and by the assistance of the eyelid, and a gush of liquid from the glands, the intruding matter is carried clear of the eye and lodged on the insensitive body developed at the corner.

Draught, or the power of drawing loads from one place to another, appears to be the purpose for which the horse is naturally adapted, owing to the centre of gravity being placed well forward, which enables the horse to utilize his own weight to assist in drawing the load.

In considering the question of a force acting on a particle there are three things to be borne in mind:—(1) the point of application (the point of attachment of the trace to the hame); (2) the direction; and (3) the magnitude of the force. In our case the last of these three things is the power of the horse, and the magnitude of this power we know.* The shoulder is the point at which the greatest power is obtained without impeding the action of any of its limbs (see Fig. 2, Plate I.), and Major Francis Dwyer, in his "Bits and Bridles, Draught and Harness," tells us that

The trace should be attached as nearly as possible opposite to the immovable part of the shoulder-blade—that is, to the centre of the shoulder-blade, which is about an inch higher than the hame-hooks of the majority of wholesale made collars.

The means of applying the power at this point by neck collars, or by breast straps, will be considered in a future chapter. The next thing is the direction of the force, or in other words the inclination of the trace. If such a thing as friction did not exist it is easily seen that the trace ought to be parallel to the road, for then the whole force of the horse would be used to draw the load along the road; but friction does exist, as we know too well, and must be overcome. It is caused by irregularities of the surface, and, in order to overcome it, the load must be lifted over these obstructions, *i.e.*, we must use part of the horse-power to draw the load along the road, and part to raise it over the obstructions. This is done by inclining the trace (see Fig. 3, Plate I.) Let A be the point of draught and A B the direction of the trace, then from B, any point in A B, draw dotted line B C perpendicular to the road, and dotted line A C parallel to the road, then we have a triangle A B C. So that if A B represents the force of the horse in magnitude and direction, then C A and B C represent the resolved parts along and perpendicular to the road. And it will be seen that, by increasing or decreasing the angle of inclination of A B, the perpendicular force B C is increased or diminished. Therefore, for a smooth road, a very slight inclination is all that is necessary, but for a rough road, a much greater inclination is required.

PLATE I.

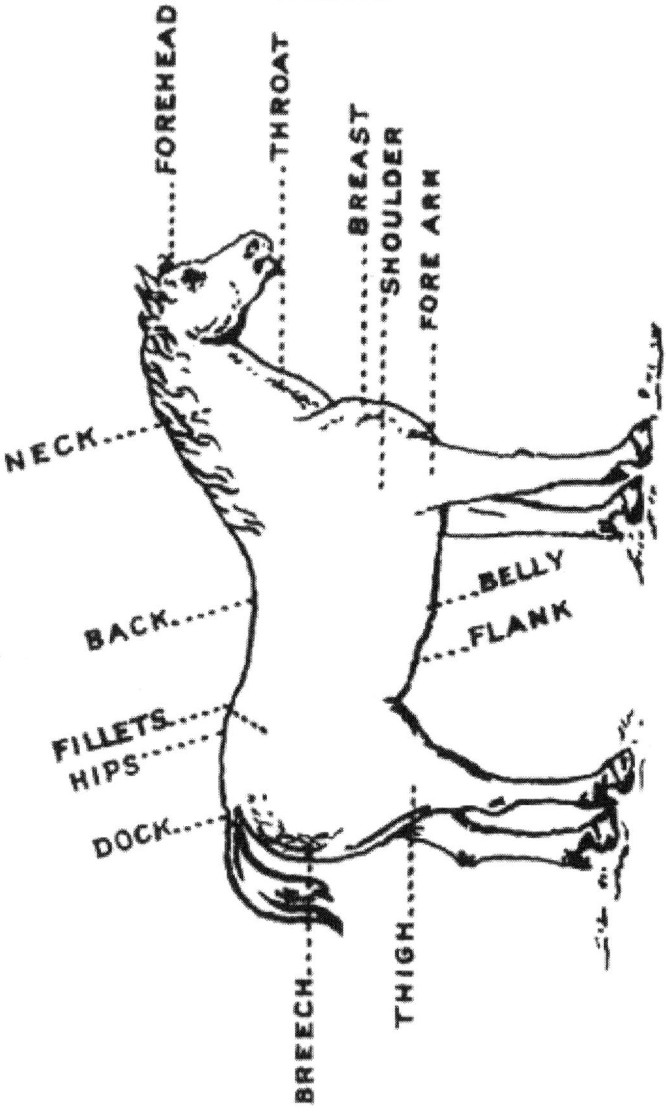
FOREHEAD
THROAT
BREAST
SHOULDER
FORE ARM
NECK
BACK
BELLY
FLANK
FILLETS
HIPS
DOCK
BREECH
THIGH

FIG. 1.

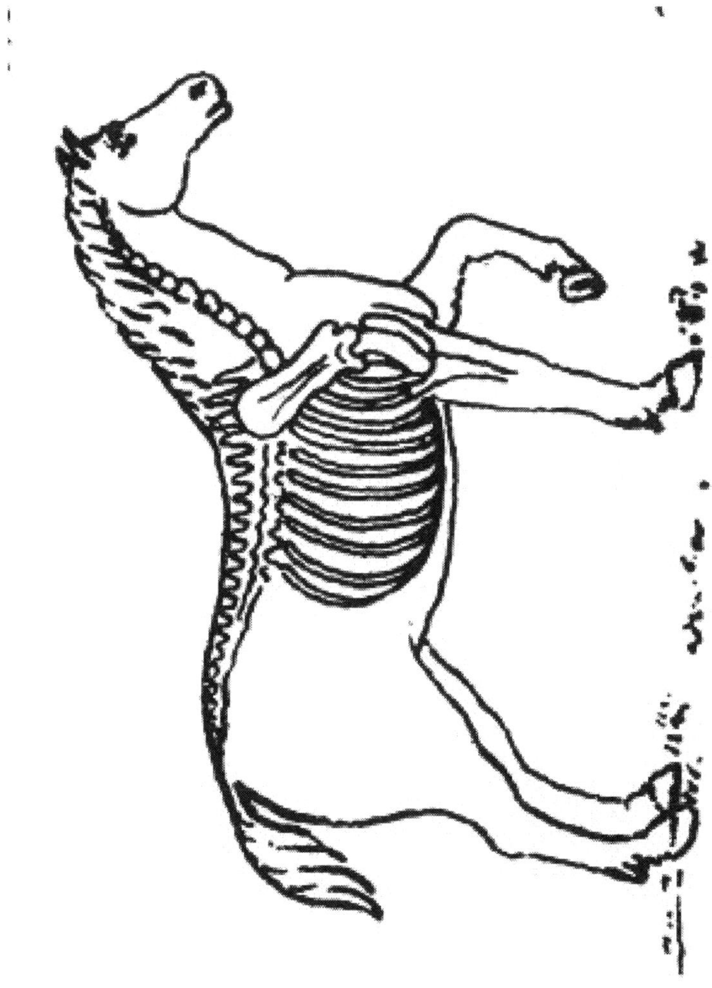

FIG. 2.

FIG. 3.

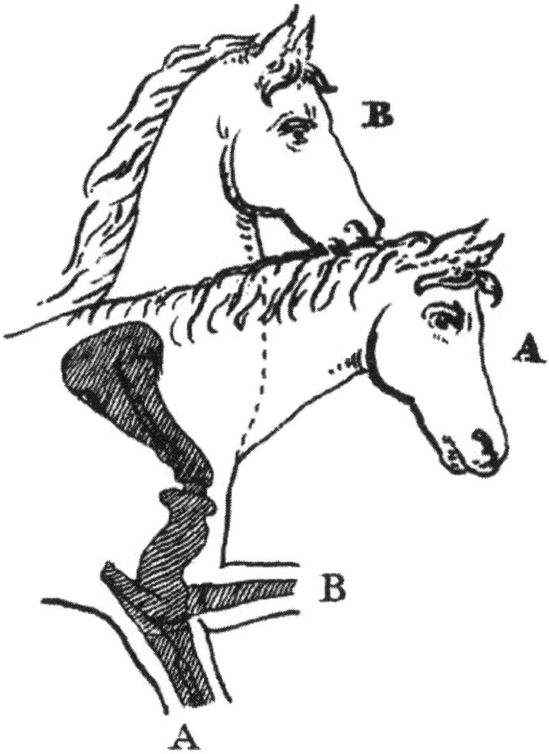

FIG. 4.

There is a muscular connection between the shoulder and head which, by this means, may be raised or bent towards the chest. "Mayhew's Horse Management" contains some very able remarks on the shoulder which I am induced to quote. They are as follows:—

In the metropolis of the world it is indeed a rare sight to behold a carriage drawn by a pair of really good animals.

The quadrupeds in general use for such purposes are mostly faulty about the shoulders. The fore-hand is placed upon the trunk in too upright a position. The job-master is conscious of this defect. He always endeavours to convince his patrons that such a make is advantageous where a creature is designed for harness. Possibly the tradesman might succeed in persuading his customers into a false belief, were not prejudice opposed to his suggestions. Ladies admire high action in the steeds attached to their vehicles; this is the kind of step which most of the horses just described are incapable of long exhibiting.

Art, or cruelty, however, can partially amend the faulty motion of the limbs. Force the head into an unnatural attitude by the unscrupulous employment of the bit or of the bearing rein; retain the neck erect without regard to the cramp induced, or heed of the strain cast upon the muscles and the torture, although the life be shortened, and the safety of the owner endangered, nevertheless, may occasion the feet to be raised during progression. This fact is illustrated in the previous engraving (Fig. 4, Plate I.) The natural mode is indicated by the letters AA, the possible change of form is to be seen in the parts distinguished as BB, although the action there depicted certainly displays a most unusual degree of amendment, to induce which must shorten the existence.

Any such improvement is always procured at a vast personal risk; for the head being raised partially throws the eyes out of use. It also impedes the circulation, ruins the mouth, distorts the body, and deranges the breathing. All these evils are inflicted to obtain the kind of pace which is never natural, but which closely resembles the sort of step that is characteristic of blindness in the horse.

The shoulder blade is much more fully developed in some horses than in others, and, therefore, necessitates a suitable collar; *i.e.*, a collar shorter at the draught than would otherwise be advisable.

If gentlemen, or their grooms, would only see that the neck collars of their horses fitted them truly, there would be little risk of sore shoulders, and probably very little need for false collars. To be perfect, a neck collar must have a prominence wherever we find a depression in the shoulder, and a depression wherever there is a prominence. By these means, and these only, can sore shoulders be avoided, and comfort secured to the horses. There are also one or two further points about the collar to which I will refer hereafter.

The same rule holds good with regard to saddles, and buyers should ascertain that they are made on suitable trees for their horses backs, otherwise sore backs, and, consequently, more suffering is the result.

On the mouth of a horse may be said to depend its value for riding and driving purposes. This important but sadly abused organ is, in the animal's tender years, of an excessively sensitive character, and the preservation of this sensibility, so far as nature will admit, should be the object of the breakers' and trainers' most especial care. If this were the case we should hear very little of "hard pullers" and "borers," and their coessential evils. There would be little need of those torturing bits, some of which are most cruelly devised to act on the upper jaw, some on the lower jaw, and others on that very tender part of the mouth, the root of the tongue. That there should be any necessity for such bits must be regretted by every humane and right thinking man, and the time when breakers and trainers are brought to realise the evil effects of harsh and severe handling will be welcomed by all lovers of man's noblest servant.

There are several old works by Philip Astley, riding master, late of H.M. Royal Light Dragoons, the dates and editions of which vary from 1775 to 1804. In that of 1775 there is very much less detail about bits than we might have expected. He commends the snaffle for hunting, and that is pretty much all he says. Certainly his dictum that "mad men and mad horses never will agree together," is not only a quaint way of putting it, but very true. The fifth edition of his "System of Equestrian Education," (circa, 1801), is much more elaborate, but still leaves much to be said. In Chapter V. he insists very much on the curb being loose and easy, and when you attempt to ride a horse in a new bit, he says you should put it in his mouth three or four mornings previous to mounting him. In the 1804 edition, Astley refers a good deal more to harness, saying:—

I would particularly advise that the horse be not oppressed with the pinching of the bit, the tightness of the harness, nor the too great weight of the carriage. . . . If you have two horses to your carriage, be sure to choose them of equal strength and spirit, otherwise the more vigorous one will be liable to be spoiled from his having the greater share of the labour in drawing, arising from his superiority of spirit.

Astley was evidently quite before us in the use, now so common, of snaffles for driving, for he says in this work, p. 25:—

My method is to ride with a snaffle and running-rein bridle; indeed I use it for my *chaise and coach horses*, having some aversion to the long wrench bit; but (he adds) the use of this kind of bridle I consider as similar to poison in the hands of an ignorant physician, for without a proper knowledge of its tendencies, nothing but destruction will ensue!

It was about this date that Van Amburgh, who made the training and harnessing of horses a study, created so much surprise and admiration by driving eight cream coloured horses through London and the provinces.

* There are 36 steps, and they are very steep.—J. P.

*Tractive power of horse 166 lbs. at rate of 2 miles per hour. Tractive power of horse 41 lbs. at rate of 5 miles per hour.

CHAPTER II.

ANCIENT AND MODERN METHODS OF HARNESSING.

I HAVE selected a few illustrations calculated to convey an idea of the manner in which the ancients harnessed their horses. An air of simplicity pervades all harness or trappings thus represented, whether Egyptian, Assyrian, Greek, or Roman, and we would do well to practice this simplicity a little more widely than we do, and thus show some consideration for the natural feelings of the animals that serve us.

In Rawlinson's Ancient Monarchies we learn that the Assyrians drove their horses by two reins, one attached to either end of the bit in the ordinary manner, and each passed through a ring or loop in the harness, whereby the rein was kept down and a stronger purchase secured to the driver. I append a paragraph on ancient bits, taken from the same book:—

The shape of the bit within the mouth, if we may judge by the single instance of an actual bit which remains to us, bore a near resemblance to the modern snaffle (see Plate II., Fig. 1). Externally the bit was large and, in most cases, clumsy—a sort of cross-bar extending across the whole side of the horse's face, commonly resembling a double axe head or a hammer. Occasionally the shape was varied, the hatchet or hammer being replaced by forms similar to those annexed (see Plate II., Figs. 3, 4, 5, 6, and 7*). The rein seems in the early times to have been attached about midway in the cross-bar, while afterwards it became usual to attach it near the lower end. This latter arrangement was probably found to increase the power of the driver.

The method of riding adopted by the Romans was with a single rein and snaffle bit as shown in Fig. 2., Plate II. With respect to saddles the reader is referred to Fig. 8, showing the Ephippium, and here it should be observed that the term Ephippium did not, as is sometimes supposed, denote a mere horse-cloth, a skin or a flexible cover of any kind, but a saddle tree or frame of wood which, after being filled with stuffing, was covered with softer material and fastened by means of a girth. It will also be observed from this sketch that stirrups were not used, but that the horses were taught to kneel at the word of command to take up their riders. This is additional evidence to prove how well it is possible to train a horse. Although Fig. 8 was taken from the walls of a house at Herculaneum, and shows that the system was practiced there, it is worthy of remark that the same custom was common in Spain, nor is such training so rare at the present time as some people imagine. This is attested by the following extract from the *Sporting and Dramatic News*, dated July 22nd, 1882, and illustrated by Plate III.:—

PLATE II.

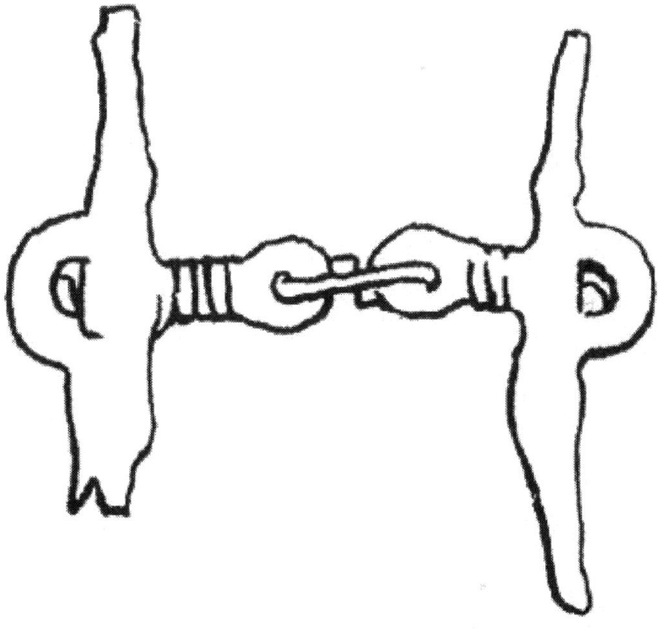

FIG. 1.

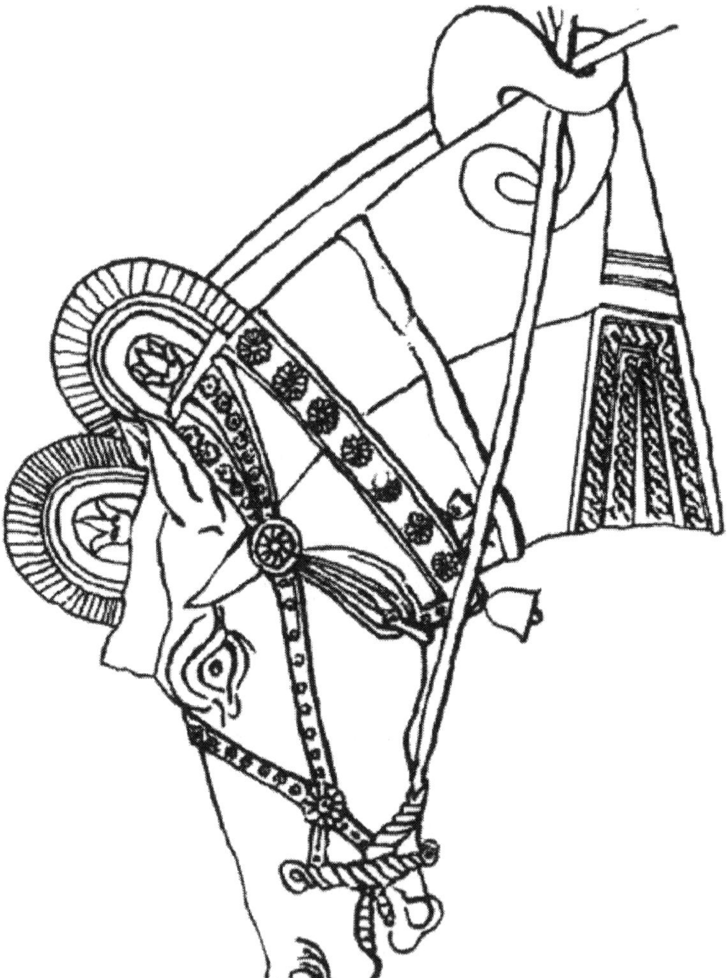

FIG. 2.

FIG. 3.

FIG. 4.

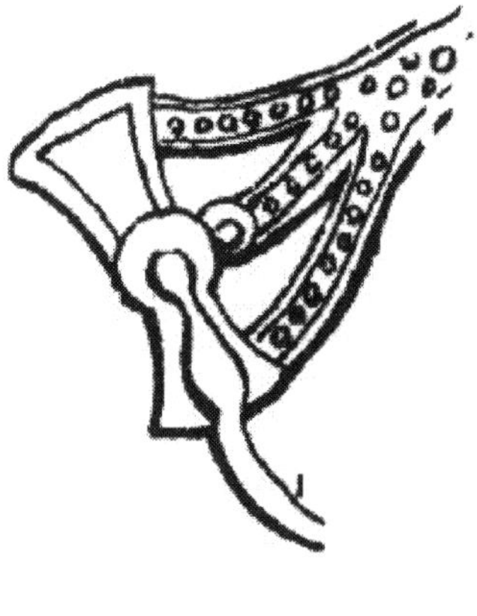

FIG. 5.

FIG. 6.

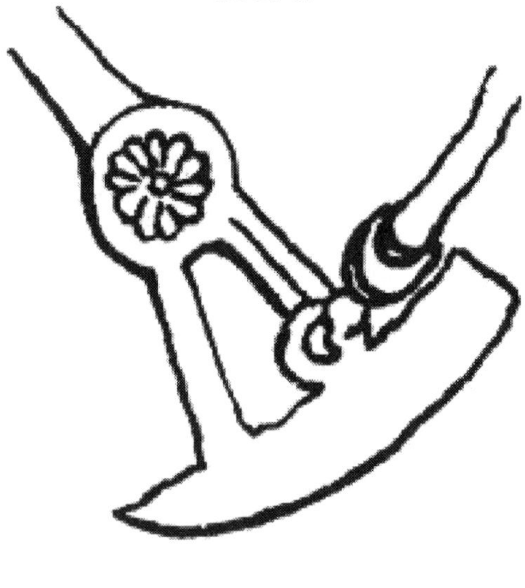

FIG. 7.

FIG. 8.

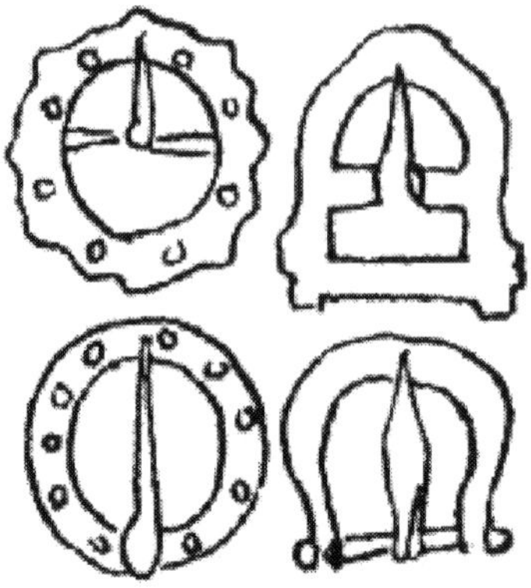

FIG. 9.

ANCIENT BITS, BUCKLES, ETC.

PLATE III.

TRAINING HORSES FOR SCOUTING (SKETCHED FORM SPORTING AND DRAMATIC NEWS)

Our illustration is from a photograph kindly sent us by an officer of the 10th Royal Hussars. The regiment has eighteen horses thoroughly trained to lie down under fire as shown. As the duty of the scout is to find the enemy, and as he often comes to grief through being suddenly exposed to fire, the advantage of having a horse obedient as a machine, trained to stand or to lie perfectly still at the will of his master, is obvious. There is, of course, also the further advantage of facilitating concealment when needed. The system which has been introduced by the 10th Royal Hussars appears worthy of imitation.

When driven with two horses the Grecian chariot was called a *Biga*, when with three a *Triga*, and with four abreast a *Quadriga*. With a Quadriga it was the custom to place the two strongest horses under the yoke and to fasten the others on each side by means of ropes. This may be gathered from Fig. 1, Plate IV., representing the chariot of Aurora, as painted on a vase found at Canosa. The reins of the two middle horses passed through rings at the extremities of the yoke. The details are more distinctly shown in Fig. 2, Plate IV., taken from a *terra cotta* found at Vienne, in France. It represents a chariot overthrown in passing the goal of the circus, the charioteer having fallen backward. The pole and yoke are thrown upward in the air, the trace horses have fallen on their knees, and the yoke horses are prancing on their hind legs. This interesting example proves that the

Grecians made their chariots so that the charioteer might regulate the weight on the withers by merely shifting his position.

The yoke by which ploughs and carriages were drawn was in many cases a straight wooden plank or pole laid upon the horses' necks, but it was commonly bent to each extremity so as to be accommodated to that part of the animal which it touched. Fig. 3, Plate IV., shows two examples of the yoke preserved in the Vatican Library. The upper drawing shows the excavations cut in the yoke with the bands of leather, the lower sketch shows the method of tying. The yoke was tied to the pole by means of a leather strap, lashed from the two opposite sides, over the juncture of the pole and the yoke.

Fig. 4, Plate IV., represents a triumphal chariot with shoulder straps attached to the pole, clearly showing the method of harnessing.

Fig. 1, Plate V., is taken from an antique Greek vase and represents a scene at the Grecian hippodrome, which is thus described:—

The goal is a mere stone post with a fillet wound round it; the form of the chariots are well shown, and the attitudes of the drivers. Each has four horses as in the earliest olympic chariot race, and the vividness of the representation is increased by the introduction of the incident of a horse having got loose from the first chariot, the driver of which strives to retain his place with the others.

Figs. 2 and 4, Plate V., show chariots of the Roman era. Fig. 3, on the same sheet, represents a mounted Roman soldier. The rein and bridle of his horse are distinctly shown, and are worthy attention.

In his admirable book, "The World on Wheels," Mr. Stratton, of New York, says:—

It would appear that the harness for curricles and that for war chariots were nearly alike, and the pole in either case was supported by a curved yoke, the end being attached to the yoke by a strong pin, bound with straps or thongs of leather to render it still more secure. The yoke, resting on a small, nicely padded saddle, was firmly fitted into a groove of metal and the saddle placed upon the horse's withers, furnished with girths and a breast band, and surmounted by an ornamental knob, in front of which a small hook secured it to the *bearing rein*. The driving reins passed through a thong or ring at the side of the saddle, and thence over the projecting extremity of the yoke, the same thong securing the girths, even appearing in some cases to have been attached to them. In war chariots a large ball, placed on the pole, projected above the saddle, which was intended either to give a greater power to the driver by enabling him to draw the reins over a groove in its centre, or was added solely for an ornamental purpose, like the fancy head-dresses of the horses, and fixed to the yoke immediately above the centre of the saddle, or rather to the head of a pin which connected the yoke to the pole.

PLATE IV.

FIG. 1.

FIG. 2.

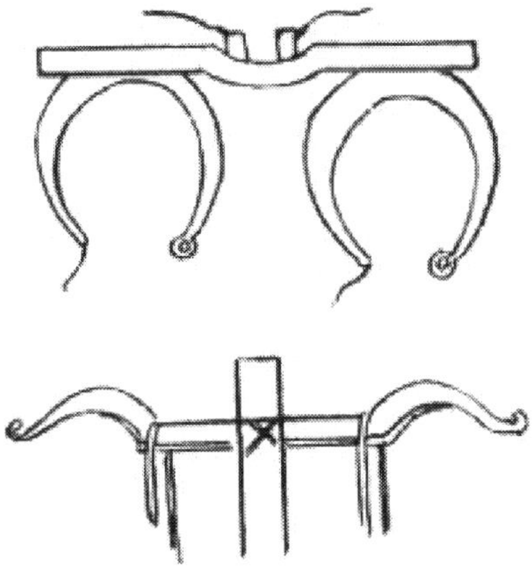

FIG. 3.

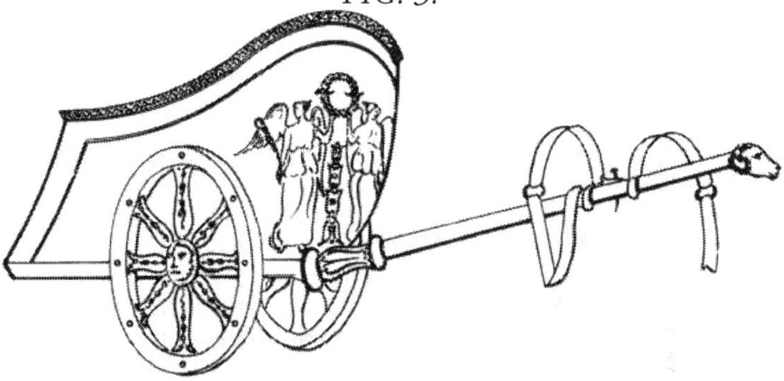

FIG. 4.

ANCIENT EXAMPLES

PLATE V.

FIG. 1.

FIG. 2.

FIG. 3.

FIG. 4

"The World on Wheels," which was published in 1878, is by far the most perfect history of carriages yet published, and I can honestly recommend it to all who take the smallest interest in matters appertaining thereto.

Did space permit I would gladly have quoted some remarks from such books as Madame Marie Isabelle's "Surfaix Cavalier on the Breaking and Training of Horses," Free Lance's "Horses and Roads," Harry Hieover's works, Reynardson's "Down the Road," and S. Sidney's "Book of the Horse;" as they all contain many valuable hints; but I purpose giving two or three extracts which I consider of exceptional interest.

In the library of the British Museum may be found a curious work by Guillaume Cavendish, Duke of Newcastle, entitled *La Methode et Invention Nouvelle de Dresser les*

Chevaux. Anvers, 1657, fol., and in the new catalogue under the heading, Cavendish (William), a work in two vols., published by Brindley in 1743, the first volume of which is the translation of the above with the original copper-plates, and the second volume a translation of a treatise on Farriery, by Boerhave and Gaspar de Saunier. Both volumes are lettered outside "Cavendish System of Horsemanship," which is clearly incorrect; but the first, by William Cavendish, is that from which I have been able to extract some observations pertinent to the subject of this work, although as a whole it is more deserving the study of the person who would qualify himself to dress or train horses for the *manège*. Although our noble author gives us a plate of rather a severe bit, he admits that a much simpler form would do as well. He says:—

It is in good lessons, well applied to the nature, spirits, and strength of every horse, that the great and subtle science of horsemanship consists, that science that can manage a horse with only a bit of wood in his mouth. The bridle I confess is of some use, though but little; art avails much more, as all your excellent riders know; for I have managed a horse with a halter only, and he went as well as with the bridle, of which I have so many witnesses in this city of Antwerp, who have seen the thing. I have also managed an English one with a scarf and made him curvet and vault very justly; so that it is not the bridle, but the art of the rider, that renders the horse tractable.

If such could be the observations of a fine horseman in 1657, how equally necessary they are in the present day when we see scores of *soi-disant* coachmen without "hands" complaining of pulling horses that would go easily enough in a snaffle. Returning to the Duke and what he says about bits, I will continue to quote him:—

There is some art in fitting a horse with a proper bridle, according as the ply of his neck is long or short, in giving him a large or small mcuthpiece, a large or narrow liberty, the eye short or long, straight or bent: the branches are either shorter or longer, stronger or weaker; mouth and liberty wider or narrower: the eye longer or shorter, straighter or more bending; the hooks according to the just measure of the bit; the curb equal, and consisting of three good round S S's, &c. . . . . But above all, this rule is chiefly to be observed, to put as little iron in your horse's mouth as possibly you can. If his tongue be very big, the liberty must be wider; if little, the liberty the less; but you must take heed that the appuy, or resting-place of the bit, be never made too near the liberty, for it would gall him, but the resting-place must be where it ought, which is about a finger's breadth from the liberty upon each side of it; and the bit must rest upon the bars, a finger's breadth above the tushes. . . . Thus you have the truth of bits brought into a narrow compass. There is little virtue in them to bring a horse to understanding, on which a man must work, and that is his reason, by the favour of the logicians distinction of reasonable and unreasonable creatures; for were they as good horsemen as scholars, they would have made another distinction. Well then, you see it is not a piece of iron can make a horse knowing, for if it were, the bit-makers would be the best horsemen; no, it is the art of appropriated lessons fitting every horse according to his nature, disposition, and strength; punishing and with good lessons rectifying his vices; rewarding him, and preserving him in his horse-virtues [this is very quaint, J. P.], which will make him a just and ready horse: and not the trusting to an ignorant piece of iron called a *bit*: for I will undertake to make a perfect horse with a cavesson without a bit, better than any man shall with his bit without a cavesson; so highly is the cavesson, when rightly used, to be esteemed. . . . The famous *Pynatel* at *Naples* never used but simple bits. . . . I always advise lenity and patience with good lessons. . . . never to exact from him above half so much as he can bear. By this means you will not only preserve him, but make him take pleasure in all that he does: he will even love you and the exercise you give him; whereas otherwise he will hate you, find the manage (*manège*) irksome, and loth to come to it.

The following is taken from an old book on horsemanship by Geoffrey Gambado, Esq., dedicated to the Right Honourable Lord Viscount Townshend, in 1737:—

I have long lamented that riding should be attended with such expense, and see no reason why many articles attending it, which cost money, might not as well be dispensed with as not. If a gentleman, when his bills are brought in, is somewhat put to it to pay for hay and corn (which, by-the-bye, are necessary, or his horses would be no better than jackasses), why should he be so very extravagant in the article of leather? Your gentlemen now-a-days must have to his bridle two head stalls and two reins; to his saddle two flaps on each side, two girths, a crupper, or a breast-plate—perhaps both—not forgetting a martingale, with its appendages. I have actually seen a young Cockader about town with a breast-plate on his horse that threw his saddle forwards, merely to be tasty, whilst his companion's nag, who slipt his girths, had a crupper on like a dragoon's. But, sir, these chaps are not confined to their expenditure in leather. Only observe the buckles and studs, &c., that ornament their trappings—not a bridle but would furnish a dozen of knives and forks, nor a saddle that has not a tea-kettle and lamp upon it.

Our forefathers never rode in this way, and I am happy that I forsee a revolution is at hand.

It is needless to reprobate such expense, as the saddler's bill probably brings in the punishment along with it.

Whether Adam, or Cain, or Abel ever rode it does not appear; but they had horses and dogs, and foxes and hares, and why are we not to suppose that they went a hunting, and had as good sport as we have (nay, better; for they had no huntsmen)? They must, however, have rode without saddles, it is pretty clear.

I do not find that these superfluities were held in much esteem in the time of the Romans. Raphael, in his picture of Attila, at Rome, has treated us with a Hun or two, riding after the fashion of their country, I suppose. The devil a bit of a bridle have they—nothing but a strap round their horse's neck, which they hold at each end with their hands.

Matters must have been somewhat changed at the end of the century, if we may judge by the following extract from William Felton's Treatise on Carriages and Harness, 1796:—

The principal properties of a harness are simplicity and sufficiency, of which there is but little variation, except in the increase of size, the ornaments or dress. Harness made for common work should not be encumbered with any superfluities, as the less a horse is burdened with trappings with the more ease and freedom he performs his work. The appearance of that noble animal, so handsome by nature, is not much, if at all, improved by dress, though it certainly is a great ornament to the equipage to have the harness display a grandeur equal with it; and one advantage is that the extra ornaments, strappings, breechings, &c., may at any time be taken off or put on at pleasure without any injury to the remaining necessary parts, so that simplicity or grandeur may at any time be preferred as he proprietors choose.

ALCESTES AND PRAETUS.—"TALES FOR YOUTH." FROM THE ORIGINAL BLOCK, ENGRAVED BY JOHN BEWICK, 1794.

* The action of these bits must have been similar to that of our Liverpool bit. Fig. 1 is evidently the forerunner of the snaffle.—J. P.

CHAPTER III.

To assist the reader who is unacquainted with the peculiarities of leather I will give a very brief *resumé* of the processes involved in the production, at such works as the large establishment of Messrs. George Angus & Co., at Newcastle-upon-Tyne, of the several kinds used for harness. To the initiated, a comparison of the methods employed by the ancients and those which chemistry has placed within our reach, furnishes many entertaining and instructive facts, but possess little interest for the general reader.

The hides of steers or heifers make the best of all harness leather when well tanned with oak bark, which imparts a light brown colour to the leather, makes it firm, with an even grain and to a certain extent allows the hide to retain its softness. The skins or hides, technically called "pelts," are first softened by being soaked in water, after which they are piled in lime-pits to loosen the hair. Instead of placing them in lime the Americans hang them in heated rooms, where they begin to putrefy, and in this state the hair is easily removed; however, neither process should be carried too far, or the grain will be injured, in which case a good finish is impossible; besides, lime is apt to make the leather harsh and brittle. Following the removal of the hair the hides are placed for a short time in a weak mixture of bark and water, termed "liquor," and from time to time they are moved into liquors gradually increased in strength until they have been in the various pits from ten to eighteen months, according to the thickness of the hides.

There are many other methods employed in tanning leather, among which we may name those known as "Hemlock" and "Union" tannage, but these and all the other numberless ways of tanning produce a rough hide, which, when curried, does not absorb the dubbing, and will not stand wear at the buckles. Moreover, this leather almost invariably swells after being exposed to rain. The reason why so much of this leather is tanned is because the process is much cheaper and is carried out in as many weeks as good bark-tanned leather takes months, but, as I have said, there is no tannage equal to oak bark. By its use we obtain the lightest, strongest, and most pliable leather. It yields a closer grained leather, which, when wet, dresses softer than any other, and to sum up its good qualities, we have only to add that it is easier to work, and takes a better and more even colour than either of the other varieties.

For the different grades of harness, such as gig, brougham, chariot, &c., a correspondingly heavy grade of leather is used in order that the traces and strapping may possess the requisite strength and substance. After tanning the hide is termed "rough leather," and from this the various kinds of leather are produced.

Merely tanning leather will not, however, fit it for the harness maker's use, and it must be subjected to another process termed "currying," by which the leather is made soft, pliable, and durable. The hide, being thoroughly soaked in water, is placed over a beam and the surplus flesh is taken off with a knife. As harness makers require their hides as level as possible, those which are naturally level, free from cuts, worm bumps, coarse necks, &c., are chosen for the best harness and rein leather, and, being exceedingly difficult to obtain, they are worth 20 per cent. more than an ordinary hide of equal weight. Shaving the hide to make it level is the most critical part of currying, as if the shaving is carried the smallest degree beyond what the hide will bear the leather is made "loose." This process is followed by the hide being scoured on both flesh and grain side in order to take out the "bloom" and dirt, by which means alone can good and regular colour be ensured. These measures also facilitate the stuffing process, which consists of the hide being well set out on a table and the application of the dubbing (a mixture of tallow and cod oil) in such proportions as the substance or thickness of the hide may necessitate. Following this, it is hung up to dry, or rather to allow the dubbing or oils to penetrate every pore, whereby the hide is rendered pliable and durable and in a state to resist wet.

It is unnecessary to trouble the reader with all the details of dressing black or brown harness leather, but, generally speaking, it consists of shaving, scouring, stuffing, and blacking, if for black harness; or staining with various ingredients if required for brown. In this state it reaches the harness maker, and he can express an opinion on its adaptability to his purpose, but no matter how wide his experience, he cannot impart to others any rules by which they can select good leather. Practice is the only thing that will enable them to accomplish it with certainty. Good leather is defined by Fitz-Gerald, of New York, "as being solid, but not hard; mellow, but not soft;" qualities that cannot be explained, but which can be detected instantly by anyone who has the requisite experience.

It is worthy of notice that the harness made by first-class firms possesses decided advantages over that produced or sold by the small dealer, insomuch that the former usually cuts out a number of harnesses at one time, and is thus enabled to use only such parts of the hide for different straps as are most suitable, such as backs for traces, backhands, and other heavy strapping, the best portion of bellies for thickening up traces, &c., and so forth; whereas the small maker with a limited demand cannot cut up his leather to the best advantage unless by keeping a disproportionate stock.

The properties of leather may be completely spoiled by injudicious cutting, and the workman who performs this operation must, above all things, have sufficient experience to enable him to so manipulate the leather that the grain shall always run parallel to the cut, otherwise the leather will stretch directly it comes into use and continue to do so.

Brown leather, which we use for reins, is in every respect the same as harness leather, except in colour, it being bleached instead of being blackened, and afterwards stained a light shade of brown. There is no part of the harness leather subject to such severe criticism as that composing the reins, and any defect in strength, cutting, or uniformity of colour may be said to spoil them. The hand parts especially should possess in great perfection the qualities of softness and pliability.

Buff leather (called by some buckskin) hand-parts are sometimes used. In making this leather the currier buffs, *i.e.*, beats out the grain, until it cannot be recognised. This gives the leather a furried appearance, devoid of gloss, like chamois skin.

Buff leather hand-parts are occasionally japanned white on one side and are then called "patent;" but when the reins are stretched by use the japan or varnish cracks, and the appearance is consequently anything but seemly.

Rein hand-parts are occasionally made of white leather, which we obtain from the hide of the horse, bleached and dressed in a solution of alum and other ingredients to preserve it. White leather hand-parts, however, are not fashionable at present, and it is more generally used for whip thongs.

Hogs' hide is made in the ordinary way from the skin of the hog, hence the name pigskin as applied to riding saddles. The most peculiar feature of this leather is that it is never made rough by friction. Imitation hogskin is made in large quantities, but as the bristles of the pig reach completely through the skin there are holes on the flesh side of the genuine article, whereas those in the imitation only reach partially through the hide. Hogskin is utilized for the pads, winkers, &c., of brown harness, which, owing to its light and showy appearance, is admirably suited for sporting carriages.

Japanned, or, as it is more familiarly known, patent leather, is made from specially selected hides tanned with more than ordinary care. After being tanned the hide is split by a machine knife, some hides yielding two and some three splits, according to the thickness of the leather, the grain side being used for enamel leather. The manufacture of patent leather includes several processes of a delicate nature, chief among them being the application of the varnish, to receive which the hide is sleeked out tightly on a board with the edges nailed down. This is the *modus operandi* at the manufactory of Messrs. Southey & Co. of London, and in other manufactories in this country, but the case is different in America where it is customary to stretch the hide tightly on screw frames with a view to gain more surface; but this method is objected to in England, as the heavy strain completely destroys the fibre and makes the leather liable to tear when stretched. The quality of patent leather is so much dependent on the state of the atmosphere when japanned that it should only be done during cold weather, otherwise it will crack on exposure to cold, and, as all consumers know, nothing looks worse than cracked patent leather.

The careful selection of leather suitable for the different parts of harness is an important consideration with the harness maker, and he is ever careful to reject any hides that have been cut or otherwise injured in the currying, or any having warbles. The presence of warbles in light strapping is sufficient to condemn it at once, especially if near the edge, when it is never safe. In addition to Fitz-Gerald's remarks, there is an excellent and very simple way of determining the tannage by examining the edge of the strap when newly cut. If well tanned the whole edge will show an uniform light shade of brown right through; but if the colour is a light hard yellow toward the centre it shows that the tannin has not penetrated, and if the edge of the strap under examination is slightly damped the difference in colour will become very distinct. Leather that has been tanned with chemicals is usually red in colour, but may be bleached to resemble oak bark tan. When chemicals are used the natural strength of the hide is usually destroyed in the tan pit, and though such leather is materially cheaper than oak tanned, it is very much less durable and will not bear the same strain. Moreover, properly tanned leather bears a bright gloss when cleaned up, very different to the harsh dry appearance of inferior tannage.

The following extract from Archæologia Æliana, November, 1861, is very interesting, as showing the esteem in which good leather was held about 100 years ago. Mr. James Clephan writes:—

Whilst I was resident in Leicestershire I accompanied some friends to Beaumanor, the seat of William Perry Herrick, Esq., on Charnwood Forest, our errand-in-chief being to see a family coach of 1740.

In the estimate from the coach-builder for the above coach there occurs the following:—

To four new harness made with the *best neats leather*, a brass plate on the edge of housing crest, housing plates, brass watering hooks, starrs, and screwd rings to ye head stalls, double bard bits, and a sett of reins, *£12*.

Furniture is a name applied to the foundations and mountings of harness, which include the hames, terrets, rings, buckles, &c. There are at present some twelve or fourteen varieties of harness mountings in use, but silver-plating in its different forms is by far the most important.

In the first instance, all buckles, terrets, pad-end loops and rings should be of solid German silver or some other white metal as a base for the close plating—a trade term, which signifies the coating of another metal with rolled sheets of virgin or pure silver by a process that would not interest the general reader.

Previous to the introduction of German silver it was customary to use cast-iron bases, toughened by a baking process, which occupied about a fortnight. White metal bases, however, do not rust and are therefore preferable to iron, which, in course of wear, shows black spots and marks by rusting through the silver.

The larger parts of plated furniture, such as hames, which must necessarily be strong, are of forged iron, neatly prepared to receive the silver by being tinned before it is applied. Those parts of the mountings subject to friction or hard wear, *e.g.* the kidney links and rings, hame pull eyes, &c., are hard plated, an extra expensive method which renders these articles almost impervious to wear.

Although there is no reliable method of determining the quality of plating, except by assay, it may be remarked that experts can detect at a glance not only the kind of plating but the grade of that kind. The amateur, however, may easily recognise hard or close plating, as the seam or joining is usually visible, especially if breathed upon or held before a sheet of white paper.

For common harness solid nickel mountings are coming into fair repute, as this metal, which is whiter than German silver, can be used unplated, and is, therefore, considerably cheaper, indeed the use of inferior leather and this class of furniture enables some dealers to sell machine-made harness at prices ranging from twenty to thirty per cent. less than can the first-class makers who use only materials of the choicest description. This fact alone will give the reader some idea of the difficulties which the best firms experience in carrying on a legitimate trade, as regards quality and price.

PLATE VI.

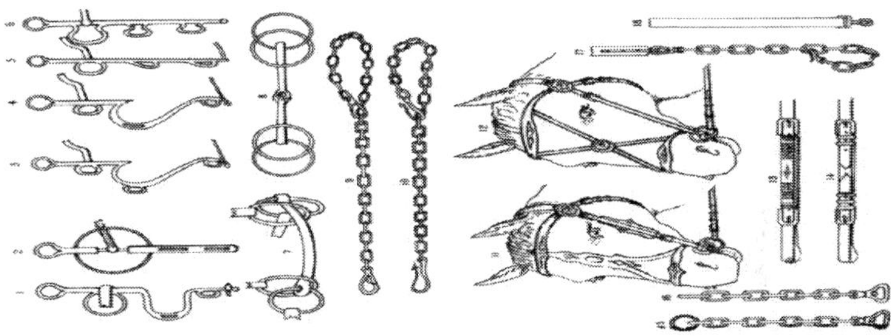

PATTERNS OF BITS, CHAINS, BRIDLES, AND STRAPPING.

PLATE VII.

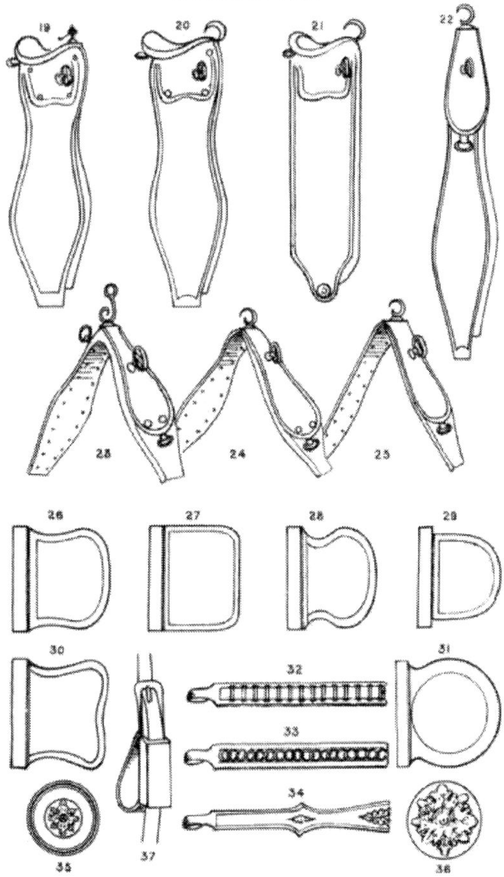

PATTERNS OF SADDLES, PADS, WINKERS. ETC.

PLATE VIII.

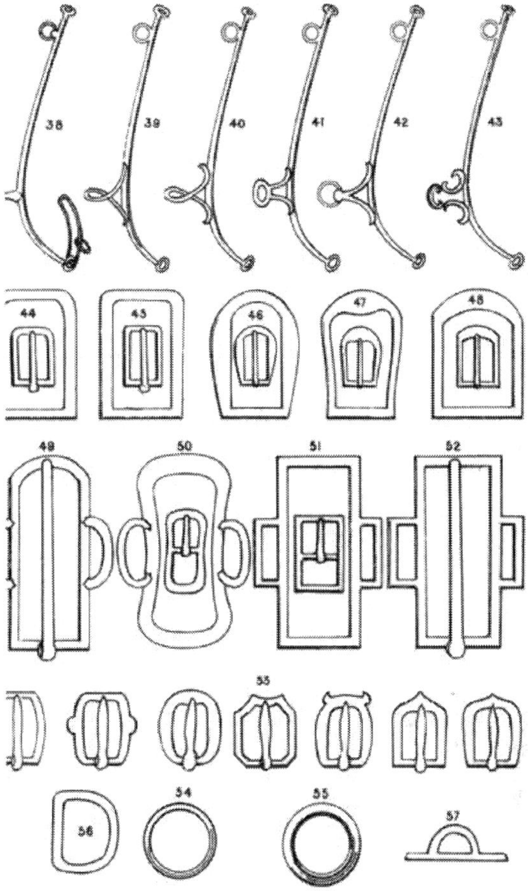

HARNESS AND OTHER FURNITURE OF VARIOUS DESIGNS.

Some firms are beginning to use a metal called nickoline, supposed to be the best of the white metals, and which it is said neither tarnishes or goes black, but on this I offer no opinion.

The buckles, &c., of brass-mounted furnitures are made of solid brass, upon which a beautifully polished surface can be produced. The hames, kidney links, and rings (see Plate VIII.), however, are made of iron and hard plated brass, in the same manner as the sheet is applied in silver plating. Many buyers prefer the terrets, hame rings, and large buckles plated, and the smaller buckles covered with leather, in which case the furniture is termed "part covered and plated." Ornaments such as crests, rosettes, &c., are raised in sterling silver, and filled with a baser metal in order to give substance.

Aluminium furniture is used in some instances, but it is costly and goes black very soon. Phosphor bronze is now offered for sale, and has the advantage of retaining its brilliancy, but beyond this feature and the durability of the metal, the writer knows very little of this furniture.

The art of covering furniture with leather has arrived at a wonderful state of perfection, and there are now methods of covering it with such compositions as celluloid, vulcanized india-rubber, &c., which give it a very neat appearance. The introduction of this style of furniture is owing to the ingenuity of some Americans, and, being durable, there is every likelihood of its coming widely into use in this country.

On Plate VI., of which I append some explanatory notes, the reader will find some patterns of the principal bits used at the present time. Bits vary greatly in price according to the material of which they are made. Cast malleable iron, wrought iron, soft steel, and hard steel are all used; but the consumer should reject anything but the very best steel bits, as, though the cost may be a little more, they are safer, much more agreeable to the horse, and can be kept bright much longer than the commoner sorts.

REFERENCE NOTES TO PLATES VI., VII., AND VIII.

1.—The Buxton pair-horse bit.

2.—Globe check or Liverpool bit.

3.—Duke's pattern.

4.—Duke's pattern.

5.—Ordinary Gig bit.

6.—Ordinary Gig bit.

7.—A Mullin-mouth Persian or Wilson snaffle.

8.—Double-ring jointed snaffle.

9.—Pole-chain, with spring hooks.

10.—Pole-chain, with ordinary hook.

11.—Bridle, without winkers; ornamental face and cheek straps.

12.—Bridle, without winkers; plain face and cheek straps.

13.—Pipe loop, showing fancy creasing.

14.—Small loops, to take strap ends or billets.

15.—Chain trace end with swivel and ring.

16.—Chain trace, with swivel and loop.

17.—Trace, with chain end and hook.

18.—Trace end, with woodcock eye.

19.—Gig saddle, with knob hook and trough for backband to slide through.

20.—Gig saddle, with bolt hook.

21.—Gig saddle, with square flaps and bolt hook.

22.—Stanhope or Imperial pad. Can be made for gig, brougham, or tandem harness.

23.—Carriage pad, with centre terret. Used for wheelers in

four-in-hand harness or pair-horse harness without leaders.

24.—Ordinary pair-horse pad.

25.—Ordinary straight side pad for lighter harness.

26.—Blinker, hatchet shape.

27.—Blinker, square pattern, with rounded corners.

28.—Blinker, another variety of hatchet pattern.

29.—Blinker, D shape.

30.—Blinker, another variety of hatchet pattern.

31.—Blinker, round pattern.

32.—Brow-band or front of flat, close-link chain pattern.

33.—Brow-band or front of curb-chain pattern.

34.—Brow-band or front of leather, fancy pattern.

35.—Bridle rosette, small pattern.

36.—Bridle rosette, large pattern.

37.—Shaft tug, ordinary.

38.—Ordinary gig hame, with anchor pull, kidney-link and ring, and jointed ring for reins.

39.—Ordinary gig hame, with branch or scroll pull.

40, 41, 42, and 43.—Varieties of same.

44, 45, 46, 47, and 48.—Gig and carriage harness buckles of various patterns.

49.—Pair-horse shoulder piece buckle, wire pattern.

50.—Pattern of buckle used for patent breeching, sometimes used throughout the harness.

51 and 52.—Other patterns.

53.—Harness buckles of various designs.

54.—Coupling ring, with round edge.

55.—Coupling ring, with bevelled edge.

56.—Pattern of Dee for breast straps.

57.—Pattern of Dee for breechings, traces, &c.

Although a simple matter in appearance, the quality and make of the thread used in stitching the harness is an important matter, and there is nothing that exhibits a plainer proof of good workmanship than tightly-drawn regular stitches, made with the best flax thread, well and carefully waxed. Sewing machines are used for the cheaper classes of harness, but for perfect finish and durability hand-made harness is unequalled, in the same sense that hand-made boots are more durable than those sewn by machinery.

CHILDREN GOING TO THE EXHIBITION.—FROM THE "ORACLES" DRAWN AND ENGRAVED BY JOHN BEWICK.

CHAPTER IV.

IT is not my intention in the following remarks to enter upon a description of the manufacture of harness, but merely to do a simple work in putting into ready and accessible form the names by which the different parts of harness are designated. Such information may be considered superfluous by those skilled in such matters, but I venture to think it will be both interesting and instructive to a numerous body who have had no opportunities of becoming practically acquainted with them.

The bridle is that part of the harness which is placed on the horse's head. Its various parts may be enumerated as follows, (see Plate IX.) viz.:—Cheek pieces [A] with buckles and loops at each end, one end taking the head-piece [B] and the other taking the bit. The head-piece [B] goes over the head, and has two loop straps by which it is buckled to the cheek and throat straps [C]. The brow-band or front [D] slides on the head-piece, and is made in a variety of ways. Many people prefer a black leather front, others incline towards coloured or striped leather; chain and beaded fronts are also used, in fact, so numerous are the available materials for fronts, and so various the shapes of blinkers, &c., that the bridle affords greater scope for the exercise of taste than any other part of the harness.

The nose band [F] slides on the bottom loop-strap of cheek-piece to which the bit is attached. The winkers [E] are generally considered part of the bridle, and are made in a variety of shapes. The principal are the hatchet, the shell, the Dee, the round, and the square patterns. Prevailing fashion is in favour of the Dee and round patterns shown on Plate VII., Figs. 27, 29, and 31, which represent them.

Bearing-reins [G] are used for supporting the head, and in the simple form are attached to a loop in the bit, from which they are carried through the loop or swivel at [H], attached to the throat band, and from thence to the terret on the pad, as shown by Figs. 1, 2, and 4, Plate IX.

PLATE IX

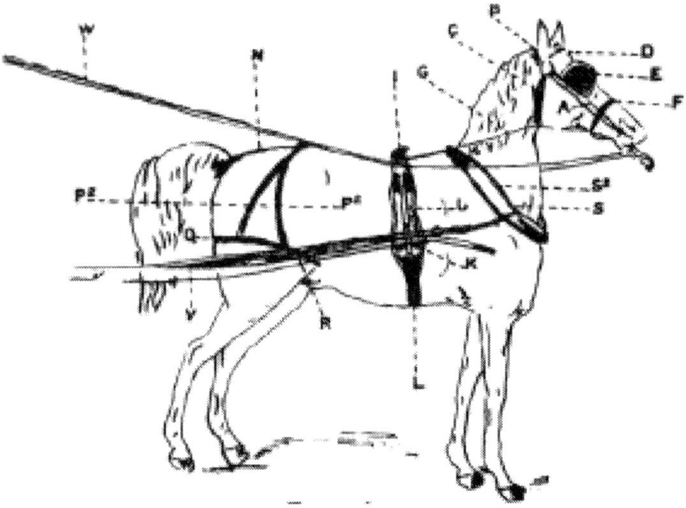

FIG. 1.

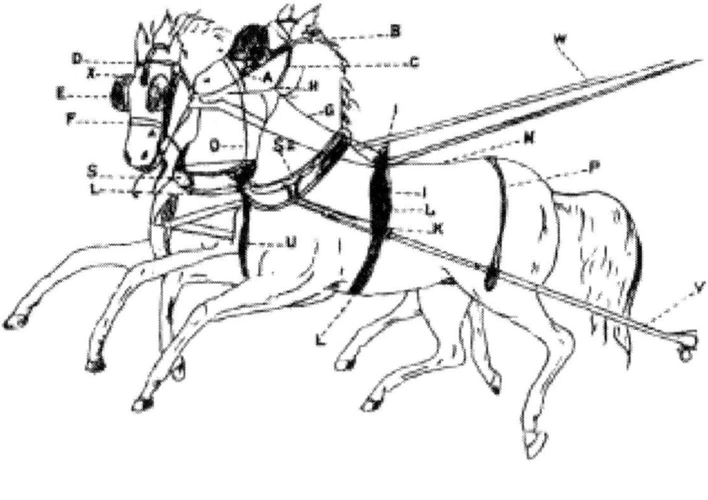

FIG. 2.

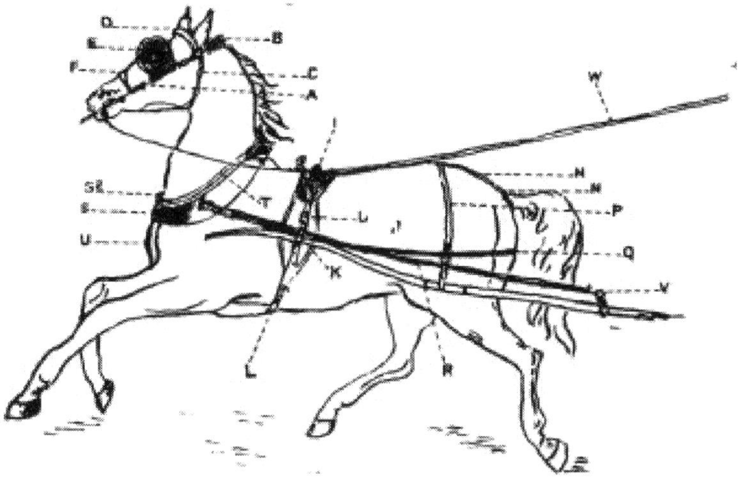

FIG. 3.

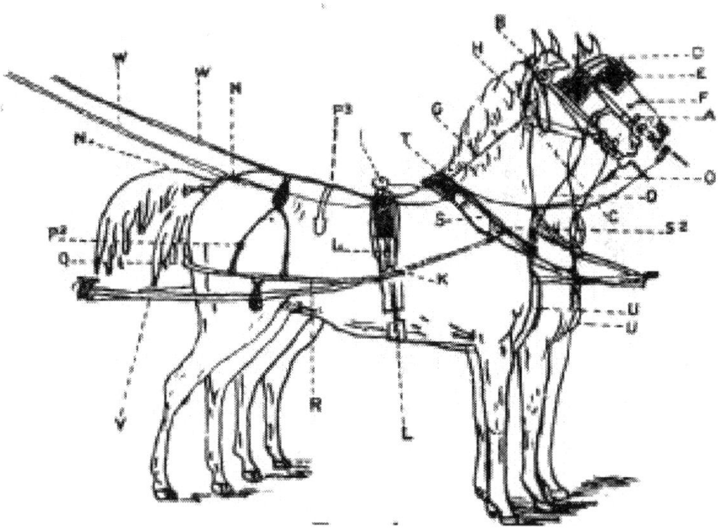

FIG. 4.

a—Cheek pieces.

b—Head piece.

c—Throat strap, lash or latch.

d—Browband or frontlet.

e—Blinker or winker.

f—Noseband.

g—Bearing rein.

h—Swivel.

i—Saddle pad.

j—Backband.

k—Shaft tugs.

l—Bellyband.

m—Pad tugs.

n—Crupper.

o—Martingale.

p—Hip strap.

*p*2—Split quarter strap.

*p*3—Fillet strap.

q—Breeching body.

r—Breeching strap.

s—Neck collar.

*s*2—Roie or roll.

t—Hames.

u—Martingales.

v—Traces.

w—Reins.

x—Forehead drop.

NAMES APPLIED TO VARIOUS PARTS OF HARNESS.

PLATE X.

FIG. 1

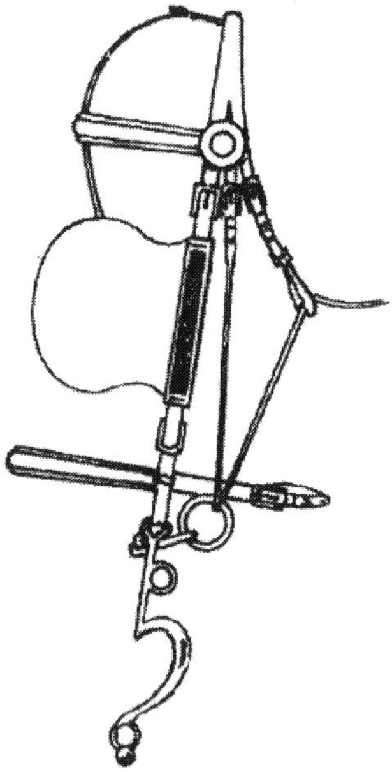

BRADOON BRIDLE.

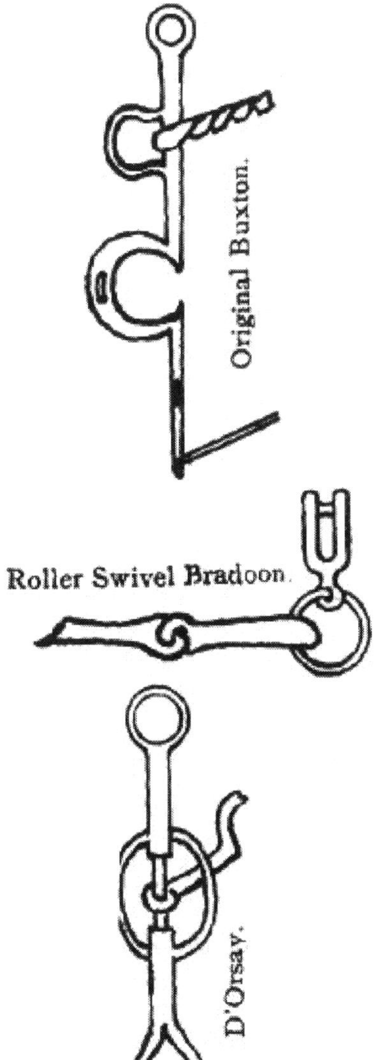

FIG. 2.

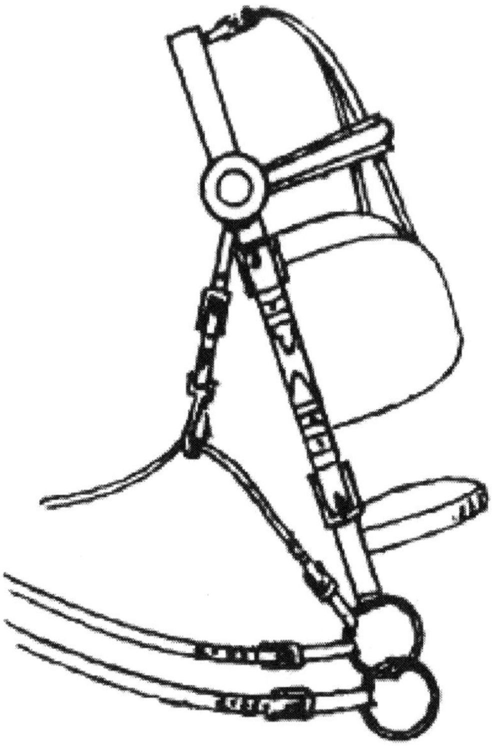

SNAFFLE BRIDLE.

MOUTH PIECES.

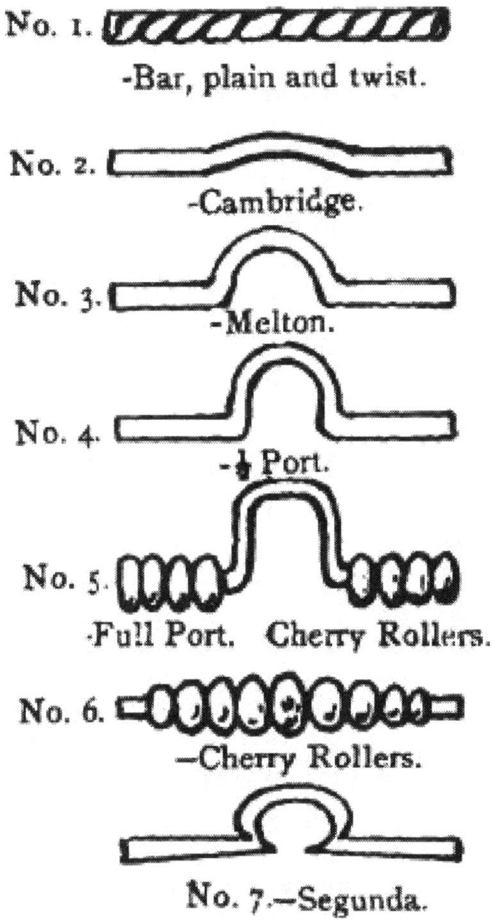
No. 1.
-Bar, plain and twist.
No. 2.
-Cambridge.
No. 3.
-Melton.
No. 4.
-½ Port.
No. 5.
-Full Port. Cherry Rollers.
No. 6.
–Cherry Rollers.
No. 7.–Segunda.

FIG. 3.

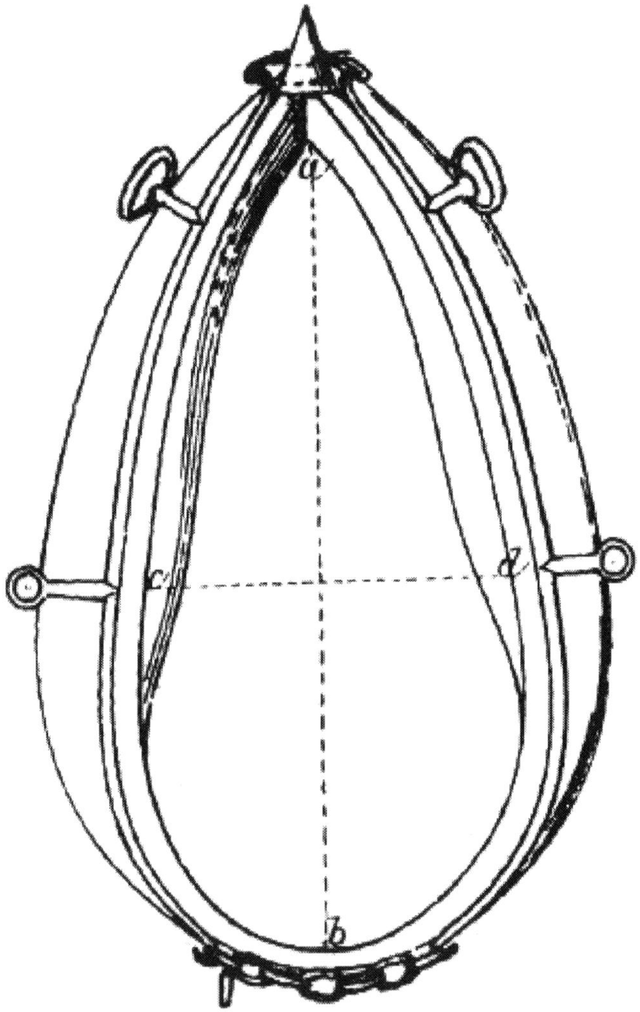

FIG. 4.

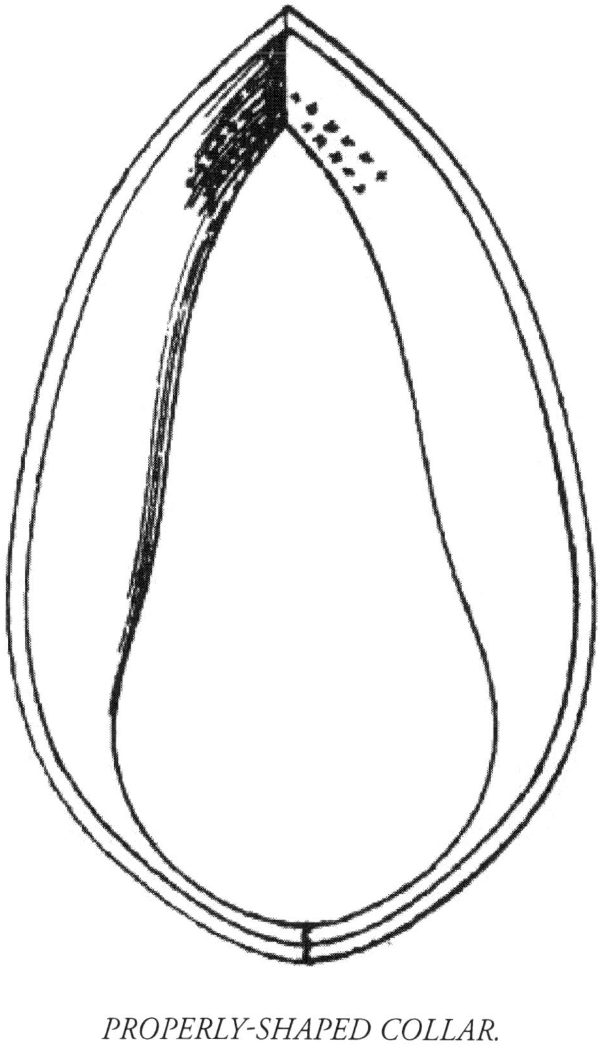

PROPERLY-SHAPED COLLAR.

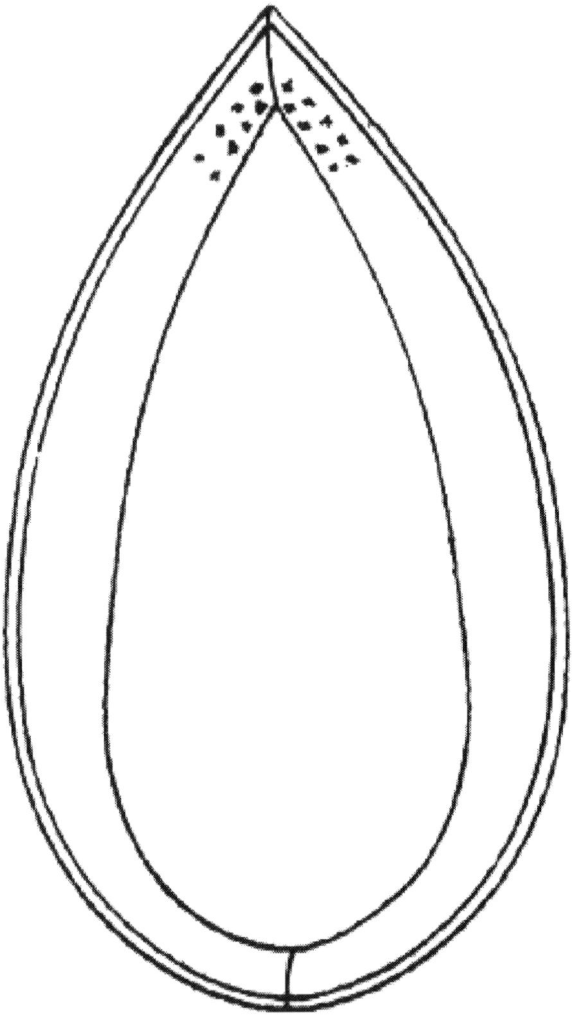

FIG. 5.

BADLY-SHAPED COLLAR.

In its more powerful form the bearing-rein is known as the gag. In this case it is attached to a billet stitched to the head-piece, and is then passed through a swivel on the bradoon bit, from which it is taken through another swivel on the end of the gag, and from thence to the terret. It is necessary that the reader should understand that two bits are used at once with the gag, one being the ordinary driving bit and the other a bradoon (see Fig. 1, Plate X.) bit, which is usually jointed in the centre and fitted with a ring or swivel, each end for attaching the bearing-rein. This is also shown in Plate X. Fig. 2 shows a bridle with simple bearing-rein.

Bridles for double harness are seldom any heavier than those used for a single horse, except in van or spring-cart harness, when the strapping is more substantial.

The saddle [I], Plate IX., resembles a gentleman's riding saddle in general outline, and is necessary for carrying the back-band [J], shaft-tugs [K], and belly band [L]. In carriage harness it is invariably covered with patent leather, but for heavy work, black oil or harness leather is to be preferred for its durability.

The pad is sometimes used instead of a saddle for single harness and is called the imperial, Fig. 22, Plate VII. It is light and elegant but much inferior to the saddle for steadiness under a weight.

The pad used for double harness has a top of patent leather, on an under cushion of oil leather. A very neat and pleasing effect is gained when the pattern of the pad harmonises with the shape of the winkers, and though this rule is pretty generally understood, there is sometimes a difference which detracts very much from the appearance of the harness. Pad cloths, which are placed under the pad, are frequently used. They are made of patent leather usually of a similar colour to the painting of carriage, and are shaped to correspond with the pad. They have a very showy appearance particularly with lace and other fancy borders.

The belly-band [L], Plate IX., in single harness is usually in one long piece, forming both belly-band and back-band, the middle and strongest portion of the strap being arranged to serve as the latter and sliding through the pad or saddle.

Double harness belly-bands are attached to the pads, the pad ends having loops to take the straps which support the shoulder-piece buckles. These latter vary greatly in design according to the class and style of the harness. (My *collaborateur*, Nimshivich, maintains that no strap or second belly-band uniting the traces is required at all.)

The pad-tugs [M], Plate IX., are made in a variety of patterns, but are usually oval in shape to admit and hold the shafts. The tilbury tug is a valuable accessory, and very convenient when a horse is difficult to "put to." It is made of iron and neatly covered with leather, and by means of a spring bolt or catch may be opened to admit the shaft. It is fitted with a loop and eye on the outside which enable the belly-band straps to be slipped through and wrapped round the shafts.

A "common" or ordinary breeching, when complete, consists of the following straps, viz., crupper [O], crupper dock [N], hip straps [P], breeching body [Q], and breeching strap [R]. Fillet straps [P3] are occasionally used.

The patent breeching, as shown on Plate IX., Fig. 3, is undoubtedly the best when kept at the proper level, of which I shall speak hereafter. It combines breeching with kicking strap, and may be attached to the shaft tug with a spring hook and eye, or with an eye bolt in the shaft.

Quarter straps, as represented on Plate IX., Fig. 2, are generally used without the breeching, and serve a two-fold purpose. They answer as kicking straps and support the trace, thus preventing the horse getting his leg over it.

The neck collar [S] is, in best work, always covered on the outside with glazed or patent leather. It is stuffed with carefully-selected straw, so disposed as to form a cushion to fit the horse's shoulder. It is strengthened with a roll [S2], of such a shape as to form a recess in which the hames are placed and secured. When this is firmly done the shape of the collar is preserved and it cannot work outside the shoulders.

Nervous or bad tempered horses sometimes require open-topped collars, which can be clasped round the neck instead of being put over the head; but such collars should be avoided, as they are neither firm nor safe. When a very soft collar is required a layer of wool may be placed over the straw.

The martingale [U] is intended to prevent the horse throwing up its head (a habit that occasionally causes some trouble). It is looped through a buckle and attached to the belly-band at one end, from whence it passes through a loop sewed on the collar, and from thence to the bit. That part of the martingale between the collar and the bit is usually a split strap with buckles on the end of each side to take the corresponding side of the bit.

Hames (sometimes called hame-sticks) [T] are necessary as a means of attachment between the traces and the collar. They are ornamental and, as I have said, prevent the collar working outside the shoulder, which it does if allowed to lose its shape.

The bullets or pulls are sometimes very fanciful in design, but the old plain pattern, known as the anchor pull, is very much in vogue just now, being strong and easily cleaned.

Shoulder pieces are attached to the hames, with buckles to alter the length of traces as may be needed, but they can be dispensed with if several holes are made in the traces at the bar. This latter arrangement does not irritate the horse's side at the bend of the shafts in the same degree as shoulder-pieces.

Traces [V], being the means of draught, must necessarily be strong, and are, therefore, made of two plies or thicknesses of the best and and soundest leather. Single traces are usually made to take the shoulder-piece buckle at one end, and have a hole at the other to take the pulling bar hook or the dart bolt on carriage. They may be fitted with a woodcock eye, but, as a rule, this is more general in leading bar traces for four-in-hands, &c.

Pair-horse traces are also made of double leather thickened up at each end to make the trace of uniform thickness, and to suit the buckles and roller bolts on splinter bar.

Leading traces for tandem or posting harness have a spring hook at the end to take an eye in the shoulder-piece buckle.

Pair-horse driving reins [W] are made both of single and double leather, the latter four times stitched, *i.e.*, there are two rows of stitching on each side, but good single leather is the most durable. It is now becoming a pretty general custom to use "flat and laid-on" reins, which are of flat, single leather, but have a narrow strip of leather laid on and sewed up the centre of the rein. Pair-horse reins consist of an outside rein attached to the outside of the bit of each horse, and two inside or coupling reins, which run from the inside of the bit of each horse, and, after crossing each other, are buckled to the outside rein at a point some eight feet from the bit. It is sometimes necessary to lengthen or shorten the coupling of the two reins, and some gentlemen prefer to have them made sufficiently long for this to be done from the driving seat, in which case the coupling is from 8 feet 9 inches to 9 feet distant from the bit.

Round reins are made with a cord running down the centre, but being unserviceable and difficult to repair are falling into disuse. The fact is, we have not succeeded so well as the Americans in the manufacture of this kind of rein. Single reins are occasionally made of double leather, stitched as before-named, but are more generally of single leather only, which is the most durable form. The hand-parts are usually of brown leather, while the body of the reins, passing through the terrets to the bit, is of black leather. All brown reins, however, are preferred by many, and are smart and attractive.

YOUNG NORTHUMBERLAND.—FROM THE ORIGINAL WOOD BLOCK, ENGRAVED BY THOMAS BEWICK, 1784.

PLATE XI

STATE HARNESS.

CHAPTER V.

PLATE XI. shows a set of dress harness, with full appointments, such as is used at present. It is complete with full bradoons, breeching with split hip and fillet straps, straight patent leather pads with pad cloths and square winkers, chain fronts, bradoon chains and breast straps. The reader will have no hesitation in deciding this style to be unsuitable for the very light carriages now used, although for old-fashioned carriages, such as are used for State purposes, it has an imposing appearance, sometimes heightened by the use of brass or silver-plated beads on the winkers, pads, &c. Plate XII. does not exhibit the same unnecessary weight. It has an ordinary bearing-rein, single hip-strap to breeching, the shoulder-pieces are drawn with safe-all, to protect the shoulders from chafing, and there are no martingales.

Leading harness for four-in-hand could be added to either of these sets, in which case the pads would be lighter or no pads at all, a diagonal strap leading from top of hame to [D] on trace, and the leading traces would be fitted with woodcock eyes to take the bars. See Plate XIV., which represents harness without breechings, bearing-reins, or pads to leaders. Plate XV. shows the style of harness generally adopted for use with the "Cape Cart." It differs principally from ours in having breast-straps instead of collars, and there are several reasons why neck-collars would be unsuitable. The breast-straps are easily regulated to suit the size of any horse, which, of course, is not the case with a neck-collar. To secure perfect ease to the riders this vehicle must be balanced so accurately that the inclination of the driver's body should take all weight off the horse's shoulder, allowing it to work easily and without imparting knee motion.

To make this clear it will be advisable to briefly explain the mode of attaching the horses. The yoke ought to be placed across the pole about 18 inches from the end. At this point the pole is covered with oil leather to reduce friction and rattle, and here the yoke should be strapped. Straps (resting on two leather pads with chapes) pass over the withers of each horse and take the yoke at the ends. This prevents the pole falling, although such a contingency is hardly likely with a well-balanced vehicle.

The breast-straps must not be hung too low, or they tend to cramp the shoulder. Each breast-strap is fitted with a [D] to take the pole braces, and by having the pole a good length the cart is prevented running on to the horses.

When this arrangement is adopted breeching and crupper dock may be dispensed with and their place taken by hip-straps to support the traces. Pads are sometimes omitted, but a light breeching then becomes necessary; this requires a light strap from the top of neck support of breast-strap to the croup, and to this strap the hip-strap is attached.

This plan is shown in Plate XVI. as devised by Colonel Pedler and. referred to in the correspondence of that gentleman with my *collaborateur*, Nimshivich.

I should go more at length here into the respective merits of the Cape Cart and the Curricle Harness were it not that the subject is fully discussed by Colonel Pedler and others in the latter part of this work, and I will therefore refer my readers to what they say. I will simply draw attention to Plate XVII., where a hansom is shown to be drawn Cape fashion by two horses, a novelty for which, in conjunction with Nimshivich, I think I may take the credit. And here I would observe that there is one characteristic about the hansom that has hardly ever been mentioned. I allude to the fact that in principle it embodies the best points of the Roman chariot in having the body and weight well over the axle and the balance adjusted to take the weight off the horse's back. So do the carts in Plates XVIII., XIX., XX., XXI., and XXII.; they all embody the principle of the old Roman chariot, especially in the method of yoking the horses.

PLATE XII

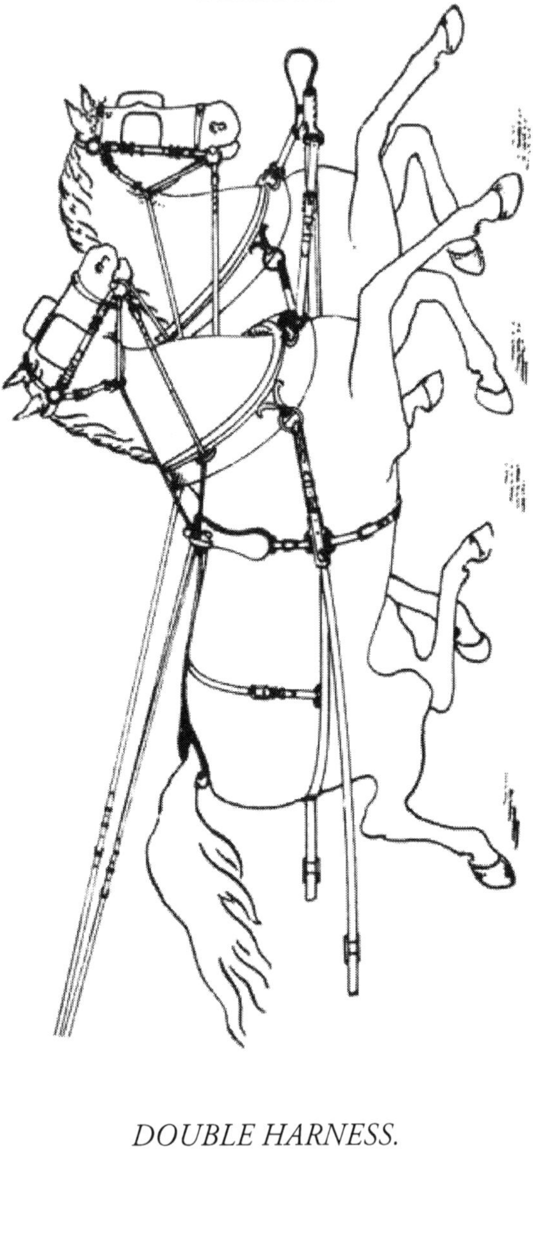

DOUBLE HARNESS.

PLATE XIII

PAIR-HORSE HARNESS WITHOUT WINKERS, BREECHINGS, OR BEARING REIN.

On Plate XIII. I have given a representation of a pair of horses harnessed in accordance with my ideas of humanity, and to this I would request the particular attention of the reader, who will ob serve that bearing-reins, winkers, and crupper docks are omitted.

Of late years there has been a great deal written about bearing-reins, but a practical experience of some thirty years with horses and harness has taught me that there are other parts of harness, in addition to bearing-reins, that may be objected to on the score of causing discomfort to the horse. It is not, however, my intention to inveigh against such appurtenances with the bitterness of some people; I would rather treat the matter with moderation, as most practical men know that there are horses and horses.

Putting aside individual caprice or the requirements of State, the harness maker, on being called on to advise his customer, has to consider appearance and utility, and, as a secondary consideration, prevailing fashion. It may fairly be averred that no strap or buckle should be present which has not a distinct use. Of course we must often admit parts the utility of which is open to dispute, and under this head winkers may be said to come. It will be observed from the sketch that I, amongst others, consider them superfluous, as they impede, and sometimes impair, the sight of an animal gifted by Providence with exceptionally powerful organs of vision. Mayhew has shown that the eye of the horse is in every way superior to that of man, and that it possesses the power of protecting itself from injury by dust, insects, &c. (a faculty, by-the-bye, that avails nothing against the lash of a whip). Why then should we use winkers at all, except for horses that have been rendered nervous and unsafe by harsh or injudicious treatment, and to prevent accidents, must be blinded by winkers, which are not only objectionable on account of their weight, but because they mar the beauty of the eye, one of the horse's best features? It is a question somewhat difficult to answer; but, on the other hand, winkers have such a body of supporters that the balance of evidence or opinion against them is not sufficient to justify their *total absence*, although I have not the slightest hesitation in asserting that they may be safely dispensed with in nine cases out of every ten. As a proof, I have only to point to the 3,000 horses employed by the Midland Railway Company, and 2,000 by the North-Eastern and London Companies. These animals are neither troubled by winkers or crupper docks, and in performing their daily work (which is as heavy as that done by any other class of horse), instead of exhibiting nervousness or unruliness, they show a degree of docility and sagacity only possible with unobstructed eyesight and perfect freedom of limb. The following is an instance:—

The Company's coach, called "The Experiment," began running on the 10th of October, 1825. One of the drivers named Dixon delights to tell of a sagacious grey that was fertile in expedients when emergencies arose. On one occasion, perceiving that a train, which had run amain, must rush into his dandy cart, he took a leap for life over the side and escaped. In a similar peril, a leap over the side being impracticable, he sprung on to the coal wagon in front, and stood like an equestrian statue on a pedestal. But the time came at last when there was no escape, and the poor old grey was destroyed.—*Gateshead Observer*, 1857.

I have said that bearing-reins have been widely condemned of late years, and this condemnation I consider deservedly merited when pointed at the gag, as this method of bearing up is both cruel and highly injurious. This remark, however, does not apply with the same force to the bearing-rein in its simple form, and the most impartial summing-up of all that has been said for and against them would not justify me in saying that they are always to be dispensed* with, for, as a writer on coaching says, bearing-reins have their uses and abuses.

In Rawlinson's "Ancient Monarchies" we learn that the use of the bearing-rein was practised by the Egyptians, but that it was either unknown to or disapproved by the Assyrians, who, in driving, used reins which were separate, not stitched or buckled together, and were held in the two hands separately. The right hand grasped the reins, whatever their number, which were attached at the horse's right cheek, while the left hand performed the same office for the remaining reins.

On going to press I have received the following from Colonel Pedler which bears most notably on the subject:—

I think we don't all of us know as much as we think on all occasions. The other day I saw a superb cart horse (and exquisite creatures they are, I think, I class them with big dogs) dragging a heavy load up the steepest hill I ever saw such a load conveyed. I spoke civilly to the carter to put him in a good humour, and then begged him to take off the horse's bearing-rein. He replied that the horse would pull himself on to his knees at once. I begged him to try, but he said he had tried, and I saw from his manner that he had, and on watching the horse I saw that he looked as if he would pull himself down. To the best of my recollection I have never driven with a bearing-rein. I was bred to look on them as wrong and cruel, and as I have seen them 99 times out of 100, I have always seen that they are so. On buying harness I have always made the bearing-reins' into pillar-reins. But I think that a loose bearing-rein, allowing a horse to put his head everywhere he can require to, but yet preventing accidents of various kinds, may be the least of two evils, but the happy medium should be preserved, and the bearing-rein not kept on an instant longer than requisite even then; and to put a bearing-rein of any sort, on many poor old cracks one sees with them, must, in my opinion, be at all times far more than ridiculous.

"Nimrod," in one of his sporting essays, said:—

There are many horses—sometimes whole teams—that will not face anything but the cheek; and where is the arm that could bear the weight of four horses leaning upon it for an hour or more together perhaps at full gallop? How much soever humanity towards horses may be enjoined, regard for our own species must prevail, and no horse in a coach or a post-chaise is safe without a bearing-rein, and for this reason, he is in constant danger from having his head at liberty, of loosing his bridle by rubbing his head against the pole or against the other horse, and then an accident is almost sure to happen.

Although I do not wish to be dogmatic about the disuse of crupper docks, I cannot help expressing my belief that they are quite unnecessary in the majority of cases, and when used in conjunction with a gag bearing-rein, are, as Paddy said, a capital arrangement for cutting a horse's tail off.

To a young horse they must be a perpetual source of annoyance, but again, I will show the reader where they are somewhat advantageous, although, for my part, I would use them as little as possible. Where it is desired that an equipage should be showy, the docks if made full will set off the horses' tails, and a double advantage arises from making them thus full—appearance and safety. For, as a horse intending to kick straight-way tucks in his tail, so by making the crupper dock very full his ability to tuck in his tail is diminished, and his disposition to kick may likely be modified. Thus, I would lay down as an axiom—use docks as little as possible, but when you do, have them made very full.

Lord Arthur Cecil has honoured me with his opinion on this subject, and I have the greatest pleasure in giving the following extract from one of his letters, not merely because it supports my views, but because of its value as the opinion of one of the most distinguished authorities of the present day on all matters appertaining to horses and driving. In speaking of carriage harness the noble writer says:—

I think that the present system, both in single and double, very nearly approaches perfection, though if horses were *carefully broken in* for the purpose I see no use in the winkers. In crowded towns only I approve of bearing-reins, so far as they make a horse careful on his feet and give the driver considerably more power in a short pull up.

Of cart harness his lordship says:—

I very strongly disapprove of *winkers*, but consider that the want of them must only be begun on young horses. Bearing-reins are ridiculous, as horses require to be able to lower their heads to the line of draught in a heavy pull. . . . Crupper docks are of no use. . . . I may say that though I disapprove of winkers, I know of exceptions in nervous horses, or even in cunning ones where the want of winkers is dangerous, in nervous ones they are frightened when they see the shafts descending on them while being yoked, and cunning ones decline to stand straight, or can see to take an opportunity of bolting when their driver is unaware, but these are only exceptions.

I have not shown a breeching, as I think the breeching unnecessary when a light modern-built carriage is used with a powerful brake. At the same time a light breeching, when properly placed, has many arguments in its favour.

PLATE XIV

FOUR-IN-HAND HARNESS.

Let us look for a moment at the arguments for and against breechings. Nimrod said:—

Many of them (the coachmen) dislike them as cumbersome to their horses, particularly in hot weather, and say further, that however useful they may be in holding back when necessary, yet they give them a trick of "sitting upon them," as they term it, whenever an opportunity offers, and not getting away from the coach when wanted to do so.

Now, if the breeching is so placed that a horse can sit upon it, he is better without it, as it is then far below its proper position, and, instead of assisting him, causes trouble by taking his legs from under him when holding back. The proper height for a breeching is the height of the hame-pulls from the ground carried round the rump, and, with a breeching in this position, the horse can hold back with perfect ease, and retain the full use of hind legs. Besides, the collar is kept off the neck, and any strain on the hame strap avoided.

Although many London omnibuses have dispensed with pads, and take the reins direct through the hame rings, they, the pads, have a distinct and useful office to fulfil. To what else but the belly-band of the pad can the martingale be attached which secures the kidney link and pole chain in case the hame strap at the top of the collar gives way? Pads, too, provide a fulcrum or lever in case of pulling horses, and, for the proper placing of reins, could not be dispensed with in the case of a four-horse team. The utility of the usual strap which connects the trace buckle with the pad, and still less of the belly-band from trace to trace, is much more questionable, and it would be much better for the leather and labour bestowed on them, to be allotted to hip straps, which are an undoubted source of safety in the case of a kicking horse. These hip-straps, as also breechings, if employed, may be sustained by a strap from the hames to the rump, in case the use of cruppers is dispensed with. The illustration (Plate XVI.) shows how this may be done, but I should prefer that the breeching, instead of being secured to a square D at A, should pass through the strap which at that point should be doubled. I show the reader thus how, for both breeching and hip straps, he may dispense with crupper dock, but, as I have previously said, I do not wish to be dogmatic about the matter.

It may be said that pads are quite unnecessary, except for unique appearance, to the leaders in a four-horse team. In their absence you place a diagonal strap from the upper side of the trace buckle to the upper part of the hames. (See Plate XIV.) I am writing, of course, on the supposition that collars are used, and not breast straps. It was in this way that Lord Arthur Somerset and Mr. Charles Hoare, than whom no better coachmen exist, harnessed their leaders in that very fast coach, the "Rapid," which ran in'78 from London to Beckenham, and the practice has been maintained, but not invariably, by the professional Selby, who then worked for them, on other coaches of his own.

The condition of the horse, so far as regards harnessing, appears to be slowly though steadily improving, which may be regarded as an indication that the barriers of prejudice are being removed, and that we are now willing to make departures from customs that have always been considered "the thing," provided there is anything to be gained by doing so; but although a remarkable unanimity of opinion does exist on the subject, there are many little matters open to discussion. The proper length of traces, for instance, is a fruitful source of dispute. Some distinguished writers are in favour of putting horses close to their work, while others equally clever hold a contrary opinion. Nimrod, in one of his letters, says:—

The traces of coach-horses should be as short as possible: they cannot be too near to their work. Those of the wheeler should just admit of his hinder leg clearing the wheel, and those of the leaders not more than sufficient to clear their tails of the bars.

I regard this as the correct view when referring to a four-wheel carriage, but with two-wheel vehicles the rule should be reversed, and the horse kept well forward of his work. Mr. G. A. Thrupp, in his able "History of Coaches," is in favour of giving horses greater license than is customary in both single and double harness, and, although I incline to Nimrod's view of the matter in the case of four-wheel carriages, I should be doing an injustice to Mr. Thrupp's good intentions did I omit to quote his remarks. They are as follows:—

I think this principle of allowing the horse greater scope for exertion is particularly worth the attention of coachbuilders. The manner in which our horses are confined by tight heavy strapping and traces, by tight pole chains, by bearing-reins and the indiscriminate use of blinkers to the bridles, has been much overdone in England. If a horse with a heavy load, driven fast over slippery roads, should stumble, it is most difficult for him to recover himself; he falls, is sometimes pushed along by the impetus of the carriage, and is more or less injured in his limbs or nerves by the accident, whilst it is difficult for him to rise again until the harness is unstrapped and the carriage is removed from above him. We also harness our horses too closely to their work in the two-wheeled carriages. We have thought only of the ease of turning and moving the whole vehicle in crowded or narrow ways, without observing the advantage of the long shafts over short shafts. If the shafts are considered as levers by which the horse supports and moves the weight behind him in a two-wheeled cart, it will be at once obvious that (although whilst these levers are parallel with the road) it does not so much signify whether they are long or short; yet the moment they point upwards and especially when they point downwards, the difference between long and short levers is felt severely by the horse. . . . . . . I think, therefore, that in future years the growth of public opinion will be in favour of longer shafts and poles.

The length of traces, however, cannot be said to affect the harness maker, as by the medium of the shoulder-piece buckles they may be lengthened or shortened at pleasure. Occasionally the buckles are fitted at the other end of the trace, in which case shoulder-piece buckles are unnecessary. This looks much lighter, and a chain end is often used to take the roller-bolt or hook, particularly in coaches and other sporting carriages (see remarks above). I have shown that draught is easiest when the trace takes a slight rise from the roller or bar to the hame pull, and this remark applies with equal force to leading traces, as nothing looks worse than a leader with traces sloping *down* to the shoulder. This is almost unavoidable when a small horse is used, and in this case long traces are to be preferred, as the point of draught can be carried further back.

Some people seem to imagine that a horse is benefited by the use of breast straps instead of neck collars, but this is a decided mistake. Breast straps are certainly light and can be utilised for any horse, no matter what his size, whereas a neck collar will only fit one; but with breast straps the point of draught is misplaced and the action of the shoulder is interfered with. A good neck collar enables us to locate the draught at its proper position, and that is much higher than is possible with breast straps; the shoulders are thus left free. By a good neck collar I mean such an one as is represented in Plate X., Fig. 4, a collar that fits close but does not nip; nor must it be slack or allowed to roll and scald the shoulders, as would the collar shown by Fig. 5, which is a fair representation of cheap work. It must be small and well laid back at the throat, but full and of good dimensions at the draught and upwards. Such a collar causes no pressure on the throat, the circulation of the blood is not impeded, and the respiratory organs are perfectly free. If the hames are made to conform to the sweep of the collar and fit closely within the roll at all points the shape is preserved and, as a consequence, the horse is comfortable. It may be asked, where are such collars to be had? And I would answer, they are very scarce and difficult to obtain except from first-class makers; but when gentlemen refuse any other collar than that which fits their horses such collars will become plentiful, but only then. Fig. 3, Plate IX., is a front view of a collar with hames connected at the bottom by a chain instead of a strap. This looks very well in traps of a sporting character.

An attentive observer will notice that our tramcar and omnibus horses (I of course refer to the horses of properly managed companies, and not to the miserable hacks that sometimes ply for hire) work more easily in neck collars than in breast straps; and, notwithstanding the hard work done by those animals, sore necks and similar evils are rather the exception than the rule.

The shape of the pad is another matter that does not generally receive the attention it deserves. If a pad is not made on a suitable tree, and stuffed to suit the shape of the horse's back (*i.e.*, to bear on the fleshy part of the back and not on the spine), there is considerable risk in its use. If too tight, the pad usually presses on the spine and causes pain, which may result in the horse trying to rid himself of the nuisance, and though he may not succeed in accomplishing this, it is within the bounds of probability that an accident happens to both carriage and driver. It was to this cause that "Nimrod" used to attribute many cases of "milling." It is the same with single-harness saddles or pads, and the buyer should make it his especial duty to see that they fit closely without being tight, and easy without rolling about, and thus necessitating a tight belly-band or girth.

There is also a point or two about the bridle, which, though in appearance slight and trivial, are not undeserving of notice. Bridles can usually be made to fit heads of various sizes by means of the straps which, in ordinary cases, are capable of being let out or taken up a few holes, but it is necessary to exercise great care in placing the cheek pieces when winkers are used, in order that the horse's eyebrows may not be chafed; it being a common practice to make winker-straps too tight to secure a close fit, which causes the winkers to chafe.

As the driver's control over his horses entirely depends on the reins and bits, buyers should select these items with more than ordinary care, especially the former. It is of the first importance that the reins be of the very best leather, whether they be single leather, round, or double leather. The greatest reliance can be placed on single leather reins, as the slightest crack or flaw on either side of the leather can be detected without trouble. As single leather reins, of the best quality, are exceedingly strong, the hand parts may be made light, and are, consequently, much more pleasing to handle, especially for a lady; but this cannot be said of double leather reins, the stitching of which allows defects to be hidden.

New reins are slippery. To obviate this, wash them well with tepid water, then allow them to dry, and taking a little powdered resin, rub it on both sides with a soft cloth. Of course this is only applied to the parts which pass through the hands.

Bits, having been so fully discussed in every journal concerned in the matter, however slightly, it would appear unnecessary to say anything about them; but it cannot be too often repeated that, in selecting bits, the buyer should be guided by the temper and disposition of the horse. A double-ringed snaffle will suit almost any horse, and if a little extra power is needed, a jointed or twisted snaffle may be used. In particular cases, the sliding-mouth bits and those having small ports may be resorted to, but they are frequently used when there is not the slightest need for them; and if the simple bits were properly placed on the bridle, so that the mouth-piece rests on the tusks, they would be found very efficacious, whereas the reverse is generally the case. If a bit is very tight, a horse cannot be expected to obey his driver, as it stops the circulation and deprives the mouth of feeling. On Plate X. I have shown several patterns of bit mouth-pieces, and would draw the reader's attention to No. 5 as a specimen of horse-torturing appliances.

In addition to the suggestions made in a previous chapter on the selection of leather, I can only add that the strapping should on no account have a greasy appearance. Double-leather strapping should be rounded or thick in the centre and tapered towards the edges; this simple precaution prevents the strap rubbing off the hair.

The best guide in forming an opinion on the workmanship is the stitching, and, although experience again steps in here, it is worth remembering that all stitching should be very regular, with the threads tightly drawn down; irregular stitching on the under-side at once stamps a trace as badly made.

These rules are probably the best that can be given on paper, for, as I have said, experience is ever the best guide; but if these points were remembered and attended to by purchasers, deception would be difficult to practice.

As silver-plated furniture is almost universally used for best harness, it is as well purchasers should be careful to have their transactions with houses of good repute, who sustain their business on the merit of their wares. The terms *best* and *superior* are words so elastic in this day, that vendors of cheap harnesses often attract customers by low priced goods, assuring them that the quality is indisputable; whereas a short period of wear proves that the cheap market is the dearest in the end. Plated goods can vary in quality twenty to twenty-five per cent., and to the casual observer the difference cannot be detected, for perhaps no other article is open to more deception than silver work. Saddlery and harness advertised as "great bargains in quality and price" should be cautiously approached.

* I quite agree with Mr. Philipson on the question of bearing-reins. Personally I have always discarded them, but it is impossible for me to ignore that jobmasters, not when they are showing horses for sale, but when they are simply exercising their cattle, use them; therefore, under certain conditions, I take them to be an advantage. Still we must remember that job-masters take care that their breaksmen have "hands," and that is just what so few gentlemen or gentlemen's coachmen possess.—NIMSHIVICH.

CHAPTER VI.

DIRECTIONS FOR THE CARE AND PRESERVATION OF HARNESS.

Whatever may be the value of the horses, the reputation of the coachbuilder and harness-maker, or the skill of the driver, a turn-out is never perfect unless it be fresh and brilliant.

The care and preservation of carriages and harness are consequently matters of importance to "connoisseurs" interested in appearance and economy, and to builders desirous of acquiring the credit due to their good work.

As few of those interested, especially of the ormer class, seem to have a knowledge of the most essential and even indispensable requisites for the proper care of harness, we trust this little work will supply a want.*

A careful coachman will pay great attention to his harness room, he will dust it frequently, will polish the wood and brass fittings, and, when opportunity offers, will take a pride in pointing out to his friends its good order and neat appearance. He will also be careful to keep everything scrupulously clean, and when brushing or polishing his harness, will select some place where cleanliness is of less importance than in the harness room. If from any cause the room is to be shut up for a lengthy period, it is advisable to cover up the harness, saddles, and all objects not placed in cupboards or cases, to protect them from dust.

GENERAL CARE OF HARNESS.—It is not absolutely necessary to clean harness, as is the case with carriages, after each time of using. Should there be no mud, nor very much dust upon the harness, it will be sufficient to brush the oil leather, to dust and wipe the japanned leather, and to rub the buckles and other mountings with a dry rag.

It is, however, necessary to wash with a sponge, but without splashing, any parts of the harness which shows signs of sweat from the horse, especially such parts as the insides of the collars, the leather panels of the saddles, the girths, &c. This washing should take place as soon as the horses have been unharnessed and rubbed down; each part being removed in its turn from the hook which it occupied whilst the horses were receiving the attention of the coachman.

This partial cleansing being terminated, the harness should be placed on trestles provided for the purpose and furnished with hooks and shelves, after being brushed the parts above-mentioned should be rubbed over with a greasy rag dipped in neat's foot oil.

WASHING.—Should the harness be muddy each part must be taken separately and washed with a sponge, special care being observed in handling the japanned leather, which must afterwards be dried with a soft damp chamois leather;

the oil leather may be wiped with an old chamois leather, that has been worn out on the carriage, the various parts are then put on a blacking up trestle, placed in the shade, and when quite dry the blacking up may commence, an essential operation after washing.

BRASS, PLATED, AND POLISHED MOUNTINGS.—Brass and plated mountings are the most difficult parts to clean; no acids or pigments of a corrosive nature should be employed, or, at least, only if used with the utmost care, and when the mountings are tarnished with verdigris. All "polishing liquids" and "metallic powders" burn or scratch the leather and varnish with which they are sure to come in contact, no matter how careful the coachman may be in their use; the remedy is apt to become worse than the disease.

Careful coachmen keep the brass or plated mountings clean by frequently and energetically rubbing them with a woollen rag, or a small chamois leather, known in France under the name of a "*peau à argenterie*," and when spotted, they use only a greasy substance, or a nearly impalpable powder, to avoid scratching. The mountings so treated should be washed with a soft sponge before again rubbing them with a dry rag.

Whiting and spirits of wine may be used with advantage for all silver-plated mountings.

The polished parts may be kept a little greasy, and, when necessary, cleaned with a burnisher or otherwise.

We would allow a little more latitude in the use of plate powder, whose effects are deleterious chiefly when brought in contact with patent leather, especially when used for arms, coronets or monograms, with which harness is frequently ornamented.

Harness makers generally supply a guard, in the shape of a piece of leather cut out to fit round the outline of the various ornaments and so protect the surrounding leather during the process of cleaning. Plate powders and acids will, nevertheless, penetrate the interstices of the ornaments, and, in spite of brushing and sponging, will gradually effect their work of destruction; the less the care exercised in their use the more rapidly will their deleterious effects be perceptible.

In some establishments a piece of flannel, moistened with lemon juice, is used to clean brass mountings; soap is applied in the same manner to silver-plated mountings, but both are used carefully and patiently.

Brushes are important factors in the production of good work; three are necessary—a hard brush to remove the dirt, a soft brush to apply the blacking, and a polishing brush.

The harness having been taken apart and brushed or washed as required, a little blacking is poured into a plate or flat vessel and applied by aid of the blacking brush to each portion of oil leather, which is then exposed to the air to dry.

As soon as it has acquired a dull or bluish tint, it is ready for the polishing brush.

Over the polish obtained by the brush, a little bees'-wax must be applied, in order to render it waterproof; a brilliancy will be thus obtained, which cannot be dimmed even by the action of rain.

The waterproof blackings are composed of bees'-wax, which is the proof, the ivory black, indigo, &c., give a colouring, and mix up into a paste. This blacking is very clean and waterproof, but the turpentine or naphtha used is not good for leather.* Prepared blackings, such as Harris' or Harding's, save time and trouble, and are as good as anything that can be procured.

BROWN LEATHER.—The preceding remarks are applicable to the black leather parts of harness; in regard to brown leather some preserve it pliable by moistening it, after it has been washed and dried, with a mixture of white wax and oil which they prepare themselves. This plan we consider good, if the leather is afterwards carefully brushed.

More frequently a brown paste of a similar composition is employed and applied in the same manner as that used for black leather.

In any case, we recommend for brown leather the use of separate polishing brushes. In fact a harness room should contain as many sets of brushes as there are kinds of leather, if the harness is to be kept in proper order.

JAPANNED LEATHER.—The japanned leather in a harness should be treated in the same manner as the japanned leather on a carriage, either by washing it with water simply, or with a little soft soap; carefully drying it afterwards with a soft chamois leather, if necessary, using a little wadding or a small piece of flannel dipped in linseed oil, to remove stains. I recommend for the patent leather on the saddles, collars, and winkers, the use of an old silk handkerchief with the feather duster. The old, soft silk will not scratch the surface and will restore the original brilliancy of the leather.

STEEL WORK.—When unharnessing the horse the bits and chains should be thrown, just as they come from the animal's mouth, into a pail of clean water where they should be allowed to remain a short time—not longer than half-an-hour. On being removed from the water they should be carefully wiped dry and, if free from rust, should be polished with the *burnisher*. If, on the contrary, any rust be apparent, it must be at once removed by the use of a little damp silver sand, very soft and fine, applied by means of small pieces or strips of old chamois leathers. When the rust has disappeared and the sand been removed by careful wiping, a *burnisher* is used to restore brilliancy.

The care of steel work generally is greatly facilitated by the use of a board fixed against the wall, and furnished with nails and hooks, to which the various articles may be attached, leaving both hands available for cleaning and polishing. This board, which we consider indispensable to any establishment of importance, is frequently mounted with sliding hooks, allowing articles of varying dimensions to be attached to them.

The same process is followed for bits, bradoons, pole ends, splinter hooks and sockets, stirrups, spurs, &c. Pole chains and curb chains are cleaned and easily polished by rubbing the links against each other in a duster or rubber, adding a little silver sand in case of rust. To clean the chains, a long bag is used, made of some stout material, sometimes of leather. This bag has a handle at each extremity, a little straw is inserted with the chains, and when, from the backward and forward movements of the bag, the straw becomes reduced to chips, the chains will have acquired a brilliant polish.

In a well kept harness room the steel work should never be greased, except in the case of a prolonged absence of the family from home. Careful coachmen will, however, during wet weather, wipe the chains with a greasy rag just before going out, to prevent rust and obviate the labour inseparable from its removal. On returning to the stables the grease is at once removed and the chains again polished.

BROWN LEATHER SADDLES, BRIDLES, &c.—Brown leather saddles, bridles, martingales, &c., are washed with a sponge and water, or if need be, with soap and a brush. To prevent them hardening we know of nothing better than good milk rubbed in with a linen rag. We recommend this application more especially for saddles; it may be rather costly but it is certainly excellent. To repolish them use a little wax or specially prepared English paste, such as Propert's or Jamieson's, applied with a brush as to other parts of the harness; to remove stains, some coachmen use a liquid called lemon water; we do not approve of this composition, it not only bleaches the leather where applied but burns and destroys it.

CARE OF VARIOUS ARTICLES.—Girths and surcingles should be washed with soap and water, and a brush if necessary; white girths should be pipe-clayed and carefully beaten when dry. Pipe-clay is also used for some kinds of apron straps, for the lashes of whips, &c. Stable halters and bridles are cleaned in the same manner as similar portions of the harness, as they happen to be of black or brown leather, &c. All covers and horse clothing, whether of tick or of wool, should be washed, and on no account put away dirty. To protect woollen materials from moth, a liberal use should be made of camphor and pepper; they should also be frequently aired and brushed. This caution is specially applicable to the woollen parts of saddles, which never spoil so surely nor so rapidly as when out of use and left to take care of themselves.

STABLE REQUISITES.—I do not enumerate here all the various tools and utensils necessary for the proper care and preservation of harness; I have mentioned the most important of them. I consider that when procuring them economy should never be studied at the sacrifice of quality. Besides requiring frequent renewal, cheap tools are generally difficult to use, and in most cases produce inferior work; we would recommend the purchase of these articles at a coachbuilder's or harness maker's, who, in addition to being competent judges of tools which are daily used in their establishments, are not likely to supply goods that will not tend to the preservation and good appearance of their own manufactures, for which they wish to obtain the credit due to good work and the increased business which a reputation for excellence is sure to bring with it.

An intimate business acquaintance of mine, who has been connected with the trades of which I am treating for a considerable number of years, has kindly contributed the following:—

Cleanliness, as it applies to harness, is an important feature as a preserver and beautifier. Leather, from its very nature, if habitually allowed to remain dirty will become hard and rigid, and more susceptible of decay, whereas by careful attention on the contrary, it retains its suppleness and softness; qualities so essential both for wear and appearance. The furniture, especially silver, deteriorates if the chamois skin and *elbow effort* is too long neglected, and here I would observe how necessary it is that proper materials should be used in cleaning furniture. The manufacturer or vendor is often blamed for what the groom ought to be held responsible for, by using those ingredients which, by their roughness, so seriously injure and destroy plate. Among the numerous preparations afloat I could advise "Goddard's Plate Powder" as a good and useful article. It can be obtained from all respectable chemists.

The same gentleman, in the following interesting remarks, carries us back a long way:—

A generation or two ago (I am writing in 1882) the usual mode of transit in Wales was on horseback, either single or by pillion, the latter being commonly adopted in attending markets or fairs. The principality could not boast of much vehicular traffic. Harness, *proper*, came into vogue with the two-wheeler gig, which in those days was built in a style we now look upon as very antique. As a commercial traveller I first took my journies in one of these gigs which supplanted the saddle-bags of my predecessor. Subsequently as trade and roads improved my gig gave place to a phæton. I am now speaking of something like forty-five years ago. I well remember how the Welsh villagers gaped at my "set out," and saluted me with halloas and amusing expressions of surprise. On one occasion I put up at a village inn known under the sign of the "Unicorn;" very humble and unpretending, but for all that the *head hotel.* The duties of "mine host" included those of the stable and the office of "John the ostler." This worthy functionary, after spending some time in adjusting the harness previous to "putting to," brought out the animal, and wheeling about, at length placed the steed's head, as the showman said (when shafted) "where his tail should be." Upon being remonstrated with, he, with good humour, so like a Welshman, exclaimed, "Why, indeed, indeed, sir, you do have such clever harnish, I do not know hardly how to use it. Indeed, we do not never see such fine harnish." Times have changed since then. The Welsh gentry and the Welsh farmer know how to appreciate a good set of harness, and the groom takes a pride in keeping it bright and nice. A pride which every true groom ought to possess. (Note.—The above was my personal experience.)

For some of my remarks on leather, harness, harness mountings, &c., in this and preceding Chapters, which accord with my own experience, I am indebted to the old established firm of Messrs. Lowe, Sleigh, Bevan, & Co., Coach and Saddlers' Ironmongers and Manufacturers, Birmingham.

MEASUREMENTS FOR HARNESS.

In ordering harness it is somewhat common to give merely the height of the horse, but to ensure a perfect fit it is essential that the horse be measured for at least two of the items, viz., the collar and the saddle. Little difficulty can exist in the case of the collar as there will in most cases be an old one to guide the purchaser, and it is therefore only necessary to give the length from A to B, and width from C to D, see Plate X., Fig. 3, (allowing 1/2 an inch to 3/4 of an inch for the shortening of the old collar). In the absence of an old collar the size round the horse's neck at the point where the collar rests should be given.

For a saddle it is necessary that the girth be given, and it will also conduce to a good fit if a rough description of the horse is supplied.

Such a description with the above particulars are usually sufficient, but, when preferred, the horses quarters may be measured for the breeching, and the throat, nose, and forehead for their respective bands. For single driving reins state the length from hand to billet, exclusive of the latter. In ordering pair-horse driving reins it is advisable to give separately the length of the reins and the length of the couplings. When possible, it is also desirable that the description of carriage for which harness is intended be given to secure proportion.

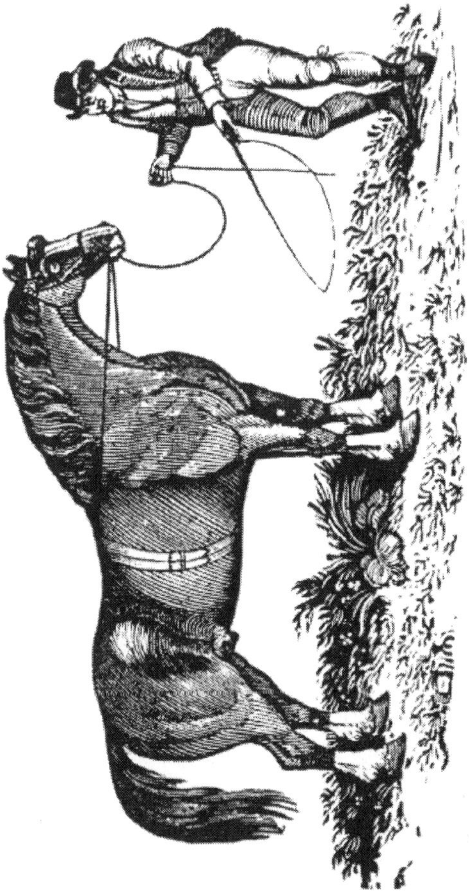

THE BLACK HORSE.—FOR DESCRIPTION SEE BEWICK'S "QUADRUPEDS," PAGE 10. FROM THE ORIGINAL WOOD BLOCK, 1800.

BIRTHPLACE OF GEORGE STEPHENSON, WYLAM-ON-TYNE.

* For the care of carriages, see Appendix.

* The most beautiful specimen of well blacked and polished harness I ever saw was in the harness room of H.R.H. the Prince of Wales, and I learnt that nothing but Everett's liquid blacking, the same as employed for shoes, was ever used. But then there was a good workman to apply it, and head coachmen who took care to see it done.

NIMSHLVICH.

THE CAPE CART.

BY

NIMSHIVICH.

ON TRACTION.

THE result of the investigations of the Dublin Society, assisted by Richard Lovell Edgeworth, F.R.S., on draught, weight, &c., of carriages more than fifty years ago, may be found in the work entitled *Edgeworth on Roads and Carriages*. The following are some of the most salient remarks in it bearing on traction:—

The higher the wheels the more advantageous is the draught *ceteris paribus* . . . . Experience has determined that the best height of wheels is from 5 feet 6 inches to 5 feet for carriages that move swiftly, and that, for heavy carriages, wheels seldom are found useful beyond the diameter of 6 feet . . . . The friction of the box of a wheel upon a well-polished axle-tree, when it is well greased, is equal to about one-fifth of the pressure of the load. So that, where the wheels are 52 inches high and the axle-trees 2 inches in diameter, the circumference of the wheel will move twenty-six times further than the inside of the box does at each revolution, but as the friction between the box and the axle-tree is only one-fifth of the pressure of the load, the actual friction will be only one-fifth of the twenty-sixth part of the load, or nearly 130th part of the weight of the carriage.

As I am only going to deal with a two-wheeled vehicle, I omit here much that was put forward in Edgeworth's work on the respective merits of high and low, and long or short coupled carriages, but the following extract from the *Carriage Builders' Art Journal* (1859), affords some pertinent observations which the purchaser of a four-wheeled carriage would do well to bear in mind.

Mr. Edgeworth, no mean authority, made this question (long or short couplings), the subject of experiment. He constructed a machine, with perch, that could be lengthened, and placed in the middle—when adjusted to 1 foot 6 inches—a weight of 28 pounds. This was drawn at the rate of five-and-a-half miles per hour, by a force of 12 lbs. The perch was then lengthened to 3 feet, and the same weight placed thereon, when it was moved at the same rate by the same amount of power. This experiment was often repeated with the same result. His conclusion, therefore, was that the draught was in the latter case no greater. But Mr. Edgeworth's experiment did not go far enough. A perfectly level, plain, and rigid model, capable of carrying ten times the weight put on, is very different to a long coupled carriage with the weight suspended between, instead of over the axles, as every strain and twist of the carriage, from the roughness of the road, is increased in proportion to the length; therefore, the shorter a carriage can be coupled, and the load placed as near as possible over the axles, the better in point of draught.

Alexander Cumming, F.R.S., who reported to the Board of Agriculture, in 1799, on the "Effects of Carriage Wheels," says:—

The several resistances that conspire to retard the progress of a carriage wheel, so far as they occur to me at present, are:—

1.—The innate force or inertia of matter.

2.—The opposition of gravity in gradual ascents.

3.—The opposition of gravity in getting over fixed obstacles.

4.—The friction of the axis.

5.—The friction or partial dragging at the rim of all wheels that are not truly cylindrical.

6.—The resistance to the rim in passing through sludge or any other substance that is partly fluid and non-elastic.

7.—The resistances of substances that have a degree of

elasticity, by which they partly recover their position when the wheel has passed, but not sufficiently strong to raise the centre of gravity of the carriage.

8th and last.—The tenacity or cohesive attraction of substances that adhere to the wheels, such as clay.

On these grounds it is maintained that a two-wheeled vehicle, with comparatively high wheels, affords the best form of traction, and that if you can attach to it a pair of horses side by side, with such harness as may secure—first, safety; second, comfort—you obtain horse-power and easy draught in their highest and best form.

It is further submitted that, with the Cape method of harnessing, these objects are attained in a more useful form than with the old-fashioned curricle. Not that for park or town this is a vehicle to be contemned, but there is nothing new about it to be pointed out to carriage buyers, or the builders they employ. It is not yet so far extinct but that any respectable builder will supply one built on the old-fashioned and, sometimes, very graceful lines. See Plate XXIII.

With these observations I will now (by permission of the Editor of the *Field*) reprint some of the correspondence that has taken place in his columns on this subject, together with some private letters which may serve to elucidate the question.

About the end of 1877, and during the spring of 1878, there might have been seen in the streets of London, at times, a cart so peculiar that it gave rise to the following letter (5th January, 1878) from the author to the *Field:*—

TWO-WHEELED CARRIAGES WITH TWO HORSES.

The other day I saw drawn up, at the Egyptian Hall, Piccadilly, a two-wheeled dog cart with two smart cobs in it, harnessed, not in curricle fashion, with the bar connecting the pads, but with what I understand Americans term a "yoke" across the pole end, and about eighteen inches from it. The ends of this yoke had straps passing over the withers of each horse, by which, at a push, the pole was prevented falling; but the general method seemed rather to be the balancing of the seats of the cart. The horses had breast-pieces, not collars, and pole chains, in addition to the yoke. An American tells me he does not remember having seen anything of the kind out there. The reason I draw attention to it is, that it seems a very practical application of the power of two horses to a two-wheeled light vehicle, which I need not point out a tandem is not. I have driven pickaxe teams hundreds of miles in old coaches and privately; but although safer than a tandem, I cannot say I like that single leader, and young men may be sure that a tandem is not the process by which to learn driving or the use of hands. Perhaps some of your readers may know something of this new rig, and supply a better description than my hurried examination permits me to do.

This letter produced several replies. All of them agreed that the querist was right in supposing that the equilibrium was maintained by balance of the seats, and in one of them it was observed that the two-wheeled carriage described was evidently the Indian tonga, common enough in the Bombay Presidency. Another writer, "Charley," was kind enough to enter more fully into details. This is a *précis* of what he says in his letter to the *Field* (12th January, 1878):—

In reply to "Nimshivich's" query, relative to two-wheeled carriages with two horses, allow me to state that I am at present, and have been for the last three years, driving a two-wheeled dogcart, with a pair of horses abreast. The cart is a Norfolk shooting cart, built by Thorn, of Norwich, with adjusting shafts, and I can very strongly recommend it as being thoroughly good; and the shafts can be removed in a couple of minutes, and the pole attached.

The pole runs underneath the whole body of the cart, and this, from past experience, 1 recommend. There is only one pin, which goes through the front of the cart into the pole. The end of the pole fits in the ordinary socket.

A lancewood bar, 5ft. long, 6in. in diameter (? circumference), is strapped on to the pole in front of the horses' breasts. This must not be fastened tightly, in order to give play to the pole. A strap goes from this bar round the collar of each horse (there should be a small piece of leather sewn on the top and sides of the collar, to prevent the strap moving on to the horses' necks). This bar must pass between the pole chains to prevent the pole from rising or falling too much.

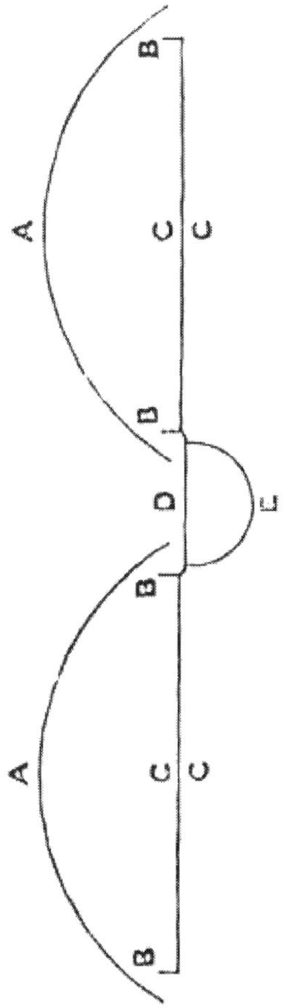

AA, STRAP TO GO OVER COLLAR; BB, BUCKLE; CC, POLE CHAIN OVER AND UNDER; D, LEATHER; E, STRAP FOR POLE WITH BUCKLE.

The above is a rough sketch, to try and explain more clearly. The leather on the centre of the bar is to prevent the pole chafing the bar. "Nimshivich" must be careful and move the steps at the side of the trap, or they will more than likely touch the horses going down hill and cause an accident, and he cannot make the pole long enough to avoid this.

The idea is not my own, but borrowed from the Cape cart. Nearly every one in the Cape of Good Hope drives a pair in a dogcart in the manner I have tried to describe. I have myself driven there for six years, and three at home, and find the yoke perfect for either rough or smooth roads. I often drive forty Irish miles in the day without distressing my horses.

Another great advantage is, if one horse falls, he cannot bring his companion down with him.

I attach the swingle trees to the place where the shafts are fixed to the body of the cart. I altered the cart myself after purchasing it from Mr. Thorn.

I need scarcely add that the chief thing is the balance. I can balance my cart so well, that merely leaning forwards or backwards, as the case may require, going up or down hill, is sufficient without moving the seat or anything else, and often take my groom, myself, and another with baggage, for a journey.

CHARLEY.

In a subsequent number of the *Field* a correspondent ("Speed") asked "What advantage this method had over the curricle bar," and "Charley" replied in the *Field* (9th February):—

In reply to "Speed." my reasons for preferring the Cape fashion, instead of a curricle bar, are as follows:—1. If horses are harnessed Cape fashion, and one falls, he will not bring down his companion, which is more than likely with a curricle bar; besides there is no chance of the bar breaking from a horse falling, which once happened to me at the Cape. 2. The pad for the curricle bar must be made heavier than the usual pad for double harness; and if the pair are not exact in pace, and, I may add, size, the bar will shift backwards and forwards—no matter how lightly it may be girthed—and the consequence is sore withers or back, and most likely both. 3. With the Cape bar, any pair may be driven, no matter as to pace or size, which is a very great advantage in a colony where one has often to drive curious pairs. Four-in-hand can be driven with a cart harnessed Cape fashion; but I fancy it would be dangerous with the curricle bar. I very often in the Cape drove four-in-hand in my dogcart. 4. A curricle bar would be a very difficult thing to get in any colony, whereas any piece of wood will do for the bar in front.

There was another letter from "Glencairn,"* in April, describing the vehicle in use in India, to which reference is made hereafter, but I will now give two other letters from the *Field* (12th January) bearing on the subject:—

TWO-WHEELED CARRIAGE WITH TWO HORSES.

The harness that your correspondent, "Nimshivich," writes about in the *Field* of January 5th is in common use throughout South Africa. Its object there is its cheapness. A collar can only fit a certain size of horse, whereas this method is self-adapting. It is also very much lighter than a collar.

Your correspondent is right in supposing that the equilibrium is maintained by balance of the seats, which are made to shift backwards or forwards.

Edinburgh, Jan. 9.

A. C. M. I.

The two-wheeled carriage seen by "Nimshivich," as described in your paper of the 5th, is evidently the Indian tonga, common enough in the Bombay Presidency.

In the ordinary tonga the yoke is padded, and rests on the neck in front of withers, with a strap round the neck, and fastened by a leather button on the end of the yoke. No traces, the whole draught being from the neck.

The best tongas I know are built at the German mission at Nassick, and cost 250 rupees (£25), with lamps, cushions, hood, English springs and axles. A pair of country "tats" (or ponies) cost from 50 rupees; a tip-top pair can be bought for 150 rupees.

UMPIRE.

This Indian tonga is described hereafter, by "Glencairn," in a private letter to me, but it evidently has nothing in common with the Cape method of harnessing.

This writer contributed the following letter to the *Field*, on the 13th April, 1878:—

TWO-WHEELED CARRIAGES WITH TWO HORSES.

"Charley," in your issue of January 9th, has conferred a boon on certain of the public, viz., those who live amongst genuine across-country roads, consisting of only two ruts—and there are many tracts of that description on Dartmoor and in Ireland. Of course two horses may be driven anywhere on the best of roads in a two-wheeled carriage, as a luxury, and a delightful trap it is. The horses in that case can travel between the wheels, and may be poled up short, as in ordinary carriage work. But I am now considering the case where a man has a two-wheeled carriage and one horse, but is unable to travel in the across-country tracks, because, although the wheels can run one in each rut, the horse cannot travel over the stones, holes, and bushes between the ruts. Roads of that description abound in wild countries, where the heavy traffic consists of carts or wagons drawn by bullocks in pairs, who go in the ruts, leaving the intermediate space untrodden; and there are many roads of the sort on Dartmoor and in Ireland, where there is very little traffic, and that little, being drawn by one horse, does not wear away the space between the ruts, because there is not enough of it, yet the wheels make two smooth ruts.

Then let the man with his dog cart and one horse exchange his horse for a pair of ponies. These are even better than a pair of horses for this purpose, because their small feet can pass each other in the narrow rut, where a large horse's foot would strike the other fetlock sometimes. A pony goes in each rut, and the wheel follows.

For this purpose the splinter bar must be as wide as the axle, and so must the yoke. They call the yoke a "neck bugle" at the Cape. And the swingle trees must be attached to the very end of the splinter bar—in other words, in front of the wheel; for otherwise the pony when in the rut would not be in front of his work. This also obliges very long pole straps, and therefore, to prevent the trap running on to the ponies, a very long pole. Half of the swingle tree projects beyond the trap, which would not do in crowded places of course, but is of no consequence on a moor.

The harness must be made with breast straps and breeching. If "Charley" will try breast straps instead of his collars, and put on breeching, he will not find his trap run on to the horses, and may replace his step. The communication between the breeching and the pole straps is perfectly direct with breast straps, but with collars breeching does not act at all. I am convinced that the reason of breast straps being universally adopted at the Cape is because they allow of breeching acting so thoroughly.

I think "Charley" will find that his yoke need not be nearly so thick. I have always seen them about an inch in diameter, and, if he will try the breast straps and breeching, I am certain he will no longer find the trap run on to the horses. Of course, breast straps are very poor draught compared to collars, but this is of little consequence in a light two-wheeled trap.

I may remark that breast straps are almost always made too broad and slung too low, thereby cramping the point of the shoulder; the breadth of two stirrup leathers is ample for a breast strap, and it should be kept well above the point of the shoulder. They require most careful adjustment. I have always had them made of the men's crossbelts, without any padding, oiling them well; the buff leather takes up plenty of oil, and is then very soft. The "D" for pole straps is fastened on by a piece of leather sewn to the centre of the breast straps. Of course, the line of sewings must be horizontal, not vertical; the latter would weaken the leather very much.

The balance is always made as even as possible by a sliding seat; but, should there be a little extra weight in front, it becomes much lessened at the end of the pole, which is a lever.

The foregoing is the Cape cart, and, if made very low like the Indian pony carts, it is, I think, the best in the world for very bad roads. Of course the cart may be made in various ways. The distinction between the Cape and Indian fashion is in the harness. The Indian fashion is equally adapted for an animal to go in each rut; and the cart is generally made very low, so as to suit the very worst roads. You sit on a platform immediately on the top of the springs, and the footboard is suspended from the platform, and hung so that it is about on a level with the axle. It should not be lower than the axle, for there are often stones, stumps of trees, &c., between the ruts, and there should be nothing to catch in obstructions which the axle can pass over.

In the Indian cart, as in the Cape cart, the ponies must be attached in a line with their work—that is, in front of the wheel—so that they can run in the ruts. The manner of harnessing the ponies varies in India, and is sometimes very barbarous; indeed, no native has the very least idea of a horse or a dog, or of anything concerning their management. None of the Indian methods of attaching the ponies resemble the Cape plan. The best is as follows:—

The pole is bent upwards in front to a level with the ponies' backs, and terminates a little in front of their withers. It runs through a yoke, which rests in grooves made in the saddle for that purpose. A pin through the end of the pole prevents the yoke from being dragged off the pole, and a pin eight inches or so further back (the yoke is between these two pins) prevents the pole from running through the yoke, and consequently stops the trap from running on to the ponies. There are no pole straps; the trap is kept back entirely by the saddles. The space between the two pins on the pole allows the yoke to travel, so that the ponies must pull by the traces, unless the arrangement is faulty; and, as the yoke revolves on the pole, one pony, in case of falling, does not drag the other down with himself. A small strap and buckle on top of the groove in the saddle prevents the yoke from jumping out, and the ends of the yoke are turned down about nine or ten inches at an angle of 45°, so that the ponies cannot possibly get further away from the pole than the breadth of the yoke, which is also the breadth of the ruts.

A spring fixed horizontally along the pole under the splinter bar takes off the knee action, which is otherwise very great indeed; the pole plays under the splinter bar on the spring, in an iron loop fixed to the splinter bar, and the end of the pole should be attached under the middle crosspiece of the body of the trap by a horizontal pin, which it can play on.

I fear the foregoing will scarcely be clear to a person who has never seen the trap. My address is with the editor, and I would gladly help any one with further particulars through the post.

The Cape cart harnessing is far the best, I think; the exceptions to perfection in it are that it must have breast straps, and that on roads where the horses are obliged to go in the ruts the pole must be very long, to enable the pole straps to be long enough to allow of the horses going in the ruts, without their also permitting the cart to run on to the horses' hocks. The Indian fashion admits of collars, and has a pole less than the length of the horses. On the other hand, the Cape fashion, from the pole being hung in front of the withers, gives no knee action, and therefore requires no spring on the pole to counteract the same.

Bombay, Feb. 20.

GLENCAIRN.

I think that this last observation, viz.:—That the Cape method gives no kind of knee action, decides the question of its superiority. On the 15th June there appeared a letter from the same well-informed writer bearing on the subject:—

I quite allow that "Nimshivich's" plan* in the *Field* of April 20th may be substituted for breeching. I adopted a modification of it myself; but I prefer the breeching because, if you keep the cart back by the saddles, the strain after all comes eventually on the crupper, and that I do not like; also the shoulders are pressed on, and that when *going down hill.* The modification which I adopted is this—and I should like "Nimshivich" to try it: First, let me say that the ordinary plan is for a strap to be fastened from each side of the saddle into the breast strap. By this arrangement the tension from point of pole to saddle is not quite direct; there is an angle where the straps from the saddle are sewn to the breast strap, and the sewing is very apt to give way. My plan is this: I sew a long strap on to a D on the inner side of the saddle (the off side of the near horse, we will say), and I sew a buckle on to the near side of the saddle, then, in putting the horses to, I pass this long strap through the pole chain D of the breast strap, and fasten it to the buckle provided for the purpose on the outer side of the saddle. The strain which ensues in going down hill is then in a perfectly direct line from the point of the pole to the saddle. I consider cruppers essential for this plan, and I do not like cruppers. I do not have any cruppers with breeching, nothing behind the hip strap which supports the breeching, and I do not even have any saddles when driving Cape-cart fashion with breeching. A light strap, like a bridle rein, goes from the top of the neck-support of the breast strap back to the croup, and to it is attached the hip strap; this is a lovely light tackle. A word about sore shoulders. I never drive with anything but a swingle bar, and that not fixed by a band of leather eight inches broad, which defeats the object of a swingle bar. but by a bolt running through it vertically, or by an eye playing on a hook. I think a *proper* swingle bar is a help to preventing sore shoulders, besides having other and most important advantages. Even in a four-wheeled carriage and with a pair I always put swingle bars, and the following is the best way to fix them, viz.: Put them on top of the splinter bar, bring an iron support from the futchells to the top of the swingle bar, and make an eye in the end of it; drive a bolt through this eye, the swingle bar, and the splinter bar, with a nut underneath. On this bolt the swingle bar revolves beautifully. You can have the ordinary roller bolts on the swingle bars. I have written to "Nimshivich," who is evidently one of those who knows what he is talking about. I consider that with swingle bars a horse is always pulling with both shoulders instead of with one at a time alternately; also that the evil of uneven traces is nullified. I consider breeching *with breast straps* to be an immense relief to horses; on coming to a descent they sit, and seem to rest, in the breeching, just keeping their legs moving, and allowing the strap to propel them along.

GLENCAIEN.

PLATE XV

CAPE HARNESS, WITH BREASTS TRAPS.

PLATE XVI

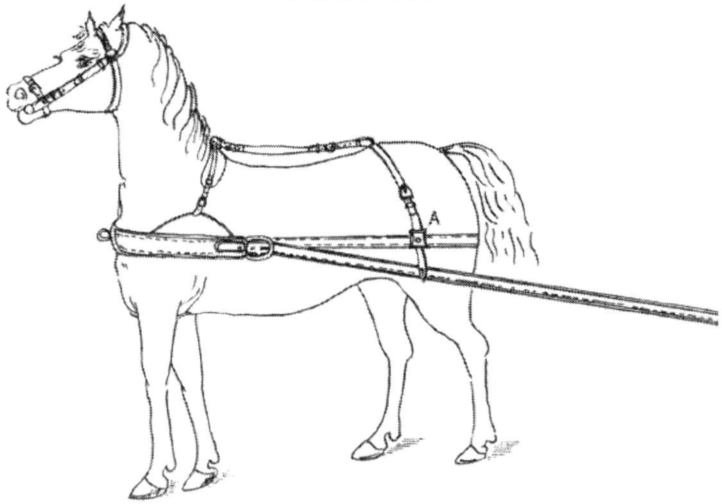

BREAST STRAP HARNESS, WITHOUT CRUPPER.

From "Glencairn" to the author, 18th May, 1878:—

I have tried to convey to you, by the enclosed (Plate XVI.), an idea of the Cape cart harness without saddle or crupper. The strap round the trace comes in very useful to prevent the breeching from getting up under the tail as I have known occur without it. Also, if a horse kicks he lifts the trace and does not kick over it.

Letter to the author from Spencer Todd, Esq., Commissioner for the Cape of Good Hope, to the Paris Exhibition:—

40, Avenue de Saffren,

Paris, 17th November, 1878.

I have your note of the 16th, and wish I could help you to a photograph or drawing of a Cape cart, but see no chance of doing so until I get back to the Cape.

The one exhibited here, now sold, was purchased by the local Commission at the suggestion. I might almost say entreaty, of Sir Bartle Frere, who alone, at the Cape, really felt how advantageous these exhibitions were.

Until very lately the only conveyance used at the Cape, that is out of Cape Town, was the Cape cart. In shape it is a dog cart, but it has a tent or hood of canvas lined with baize or cloth. The commoner carts have this hood fixed on them only in the Eastern Province, and more recently in Cape Town is this hood made, like that of a buggy, to shut down.

I have no idea of drawing, but No. 2 may give you some notion of what the common old-fashioned cart is. No. 1 is the improved reversible hood. These carts are very light. They generally hold four people, but as passenger carts they are longer and broader in the body, so as to give three seats for three persons on each; when of course they hold nine, including the driver.

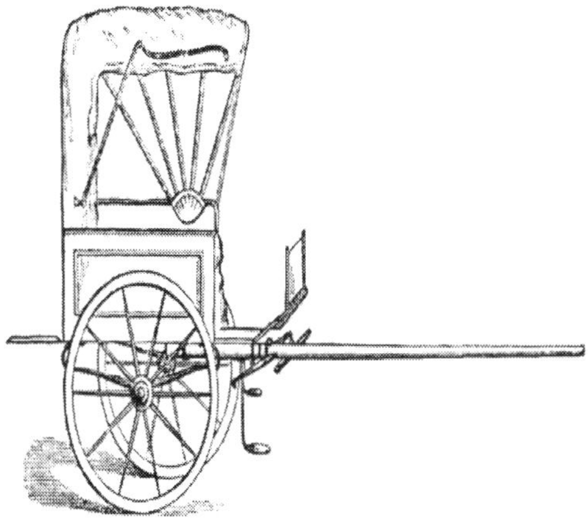

I wish the British public could know something about a capital Cape institution, for the cart is that. We don't use curricle harness but support the pole (it requires very little support when properly balanced, and that is done by shifting the front seat back or forward) by a bar of wood placed under it, in front, across the chests of the horses. Two straps pass over the necks of the horses.

PLATE XVII

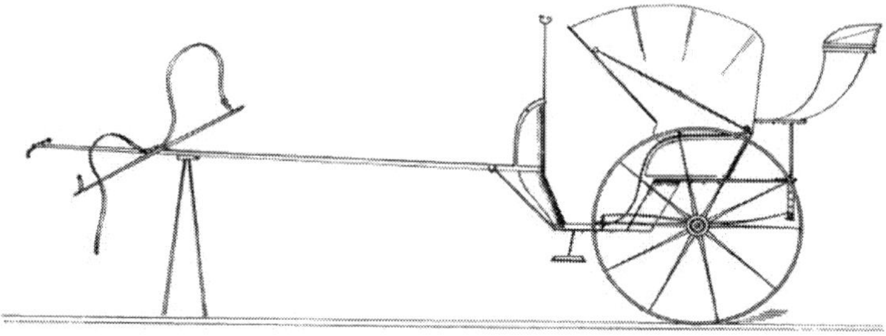

HANSOM CAB, WITH CAPE CART FITTINGS.

PLATE XVIII

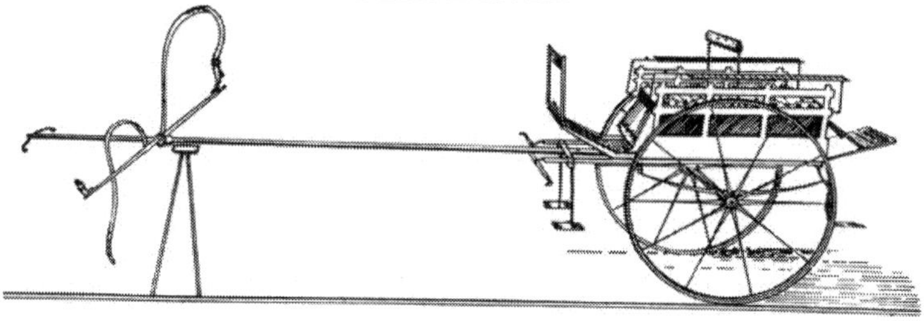

CAPE DOG CART.

I shall be in London in a few days, and should my friend, who is a member of the club whence you date your letter, be in town I may call and tell you more, being always glad to get anything that appertains to the Colony I have made my home made known.

SPENCER TODD.

"Glencairn" subsequently favoured the author with the

following sketch and description of the Indian cart pole, from which it is clear that the whole arrangement varies considerably from "Charley's," inasmuch as the latter has a straight pole and the yoke in *front* of the ponies, whilst here the yoke rests on the saddles or pads:—

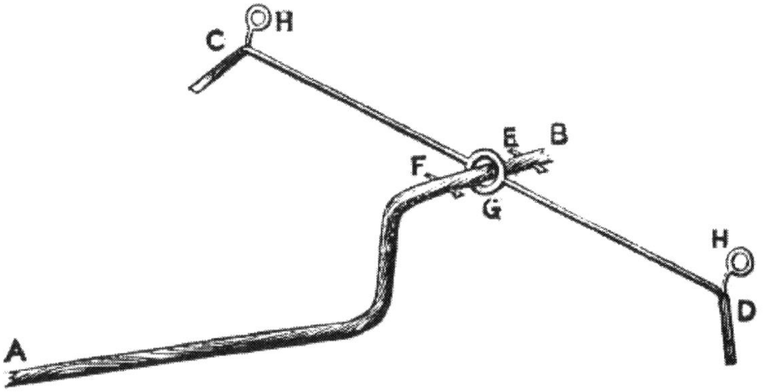

INDIAN CART POLE AND YOKE.

A B pole; C D yoke, which rests on saddles; E pin through pole which prevents yoke from being dragged off; F pin through pole which prevents cart from running on to the ponies; G centre of yoke in the form of a ring which travels on the pole from E to F, without this play it is found that the ponies are often pulling by the saddles instead of by the traces. This ring also *revolves* on the pole, so that if one pony falls the other is not dragged down; the pole is of wood, the yoke of steel; H H territts; the hand rein only is passed through a territt, the coupling rein goes *under* the yoke, and through a yoke territt in the harness to the other pony's bit. the saddles have no territts; the saddles are rather "pack saddle" looking things, they have a groove, or crutch, on top, in which the yoke rests.

N.B.—There is no territt on the outer side of the harness. The hand rein should go direct from the territt on the yoke to the bit. This gives immense power, and is the more valuable as the ponies are generally stallions who spend their time when out of harness—and sometimes when in—in fighting like bulldogs. I ought to say that this way of fixing the reins is merely a plan of my own, and that no one has, up to the present, paid me the compliment of adopting it. Every one else has four territts on the yoke.

From "Glencairn" to the author, 24th January, 1879:—

THE INDIAN CART.

I have succeeded in getting a photo or two of the "Indian cart" to send you, and hope they may be explanatory. Most people drive with breast straps, not collars. I also send a photo. of the way I arrange the Indian cart myself, and which I much prefer, because it admits of pole straps and breeching. I look on the latter as most important. My ponies really *rest* in it going down hill. They *sit in it* and just move their legs while the cart propels them. The yoke merely supports the pole by my plan, does not keep the trap back at all, as the loop on which the pole is slung can travel along the pole. After all, the only distinction from the English curricle is the yoke, turned down at the ends, instead of a regular curricle bar and the far stronger saddle. Both are a great improvement in the way of strength and safety I think.

All these pictures were taken from my own little cart, into which I can put either description of pole or yoke. I keep the ordinary pattern because, when travelling post. I find it convenient to have the one that strange stablemen understand. Also I put breast straps on those occasions for the same reason.

To give you an idea of the size of things in these photos. I may mention that the ponies are exactly thirteen hands, or rather. the near wheeler is, the other being less.

If I *do* drive mine on the Indian pattern when *not* travelling post then I employ collars and not breast straps; for if you cannot have breeching (and you cannot employ it with the Indian pole which comes to a level with the withers), I see no advantage in breast straps. By-the-way, with breast straps and breeching, I have the "Dee" for pole-pieces *not* in the centre of the breast strap, but at the point of the inner shoulder or thereabouts, the intention being to have it at the nearest point to the end of the pole. I contend that this is mechanically proper. Please do not suppose that I commend the Indian style of harnessing, for I consider the keeping the cart back by the top of the saddles pressing on the withers to be the worst I know. And I drive my cart either with my modification of an English curricle or "Cape" fashion, merely adopting the low centre of gravity for bad roads and the ponies being able to go in the ruts.

Field, 17th September, 1881:—

THE CAPE CART.

The present made by the young princes to their august parents of a genuine Cape cart may again draw attention to a useful form of vehicle where it is desired to attach two horses to one with two wheels. By the courtesy of Col. Kingscote, Master of the Horse to H.R.H., I have been permitted to inspect the above, and so far as harnessing, yoke, &c., goes, I find its arrangements are similar to those described by "Charley" at the time I first mooted the subject in your columns; but there is much greater space for ingress in front of the wheel than what was suggested by one of your correspondents at the time, who spoke of the possible necessity for removing the front step. There is quite two feet from front of wheel to splashboard. The small size of the Cape horses is well denoted by the whiffle-trees in this example, and I fancy that H.R.H. will have to lengthen these to adapt them to even moderate sized cobs at home. This cart holds six persons conveniently, the near front seat moving to permit passengers to seat themselves in the centre seats, and the light roof or canopy, with its well-arranged sides, might be copied with advantage. Solid good work, sensible tyres overlapping the felloes, and general excellence justifies that meed of praise which is due from us at home to colonial brethren for practical examples, and it should be paid to the constructor, Mr. King, of Cape Town.

PLATE XIX

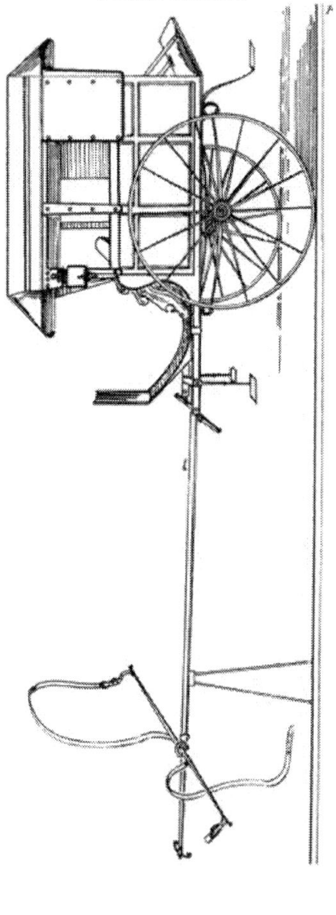

CART BUILT AT CAPE TOWN. PRESENTED BY THE YOUNG PRINCES TO THEIR ROYAL HIGHNESSES THE PRINCE AND PRINCESS OF WALES.

PLATE XX

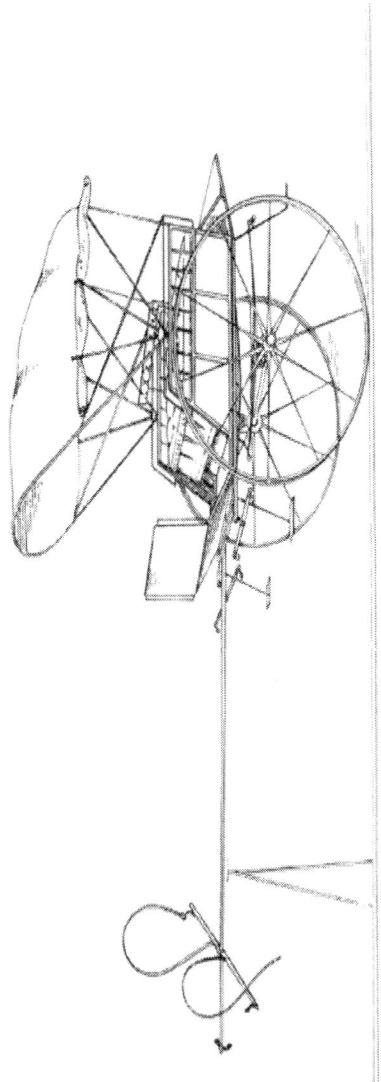

MOOR CART, WITH CAPE FITTINGS.

PLATE XXI

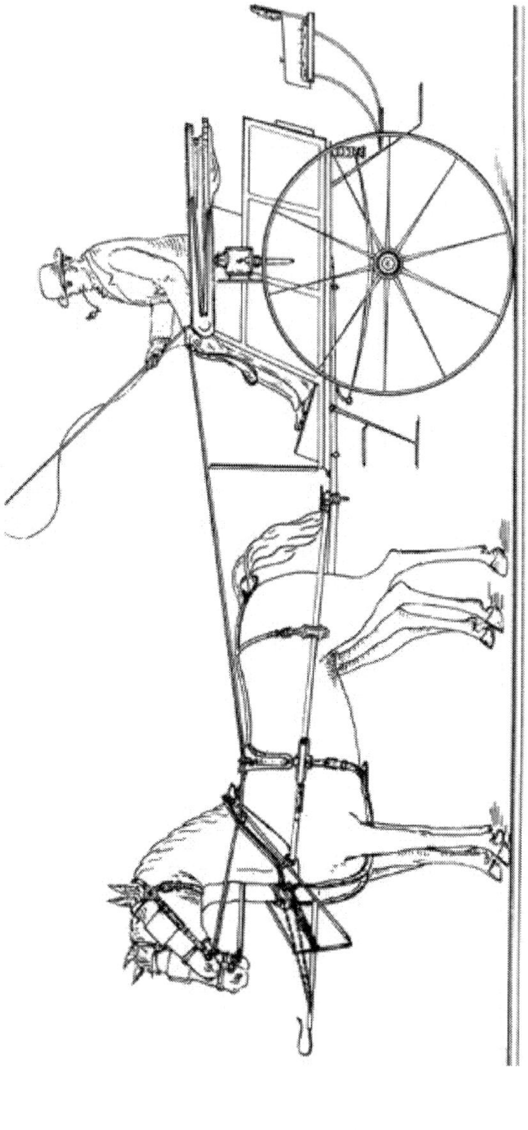

HOODED BUGGY, WITH CAPE HARNESS.

PLATE XXII

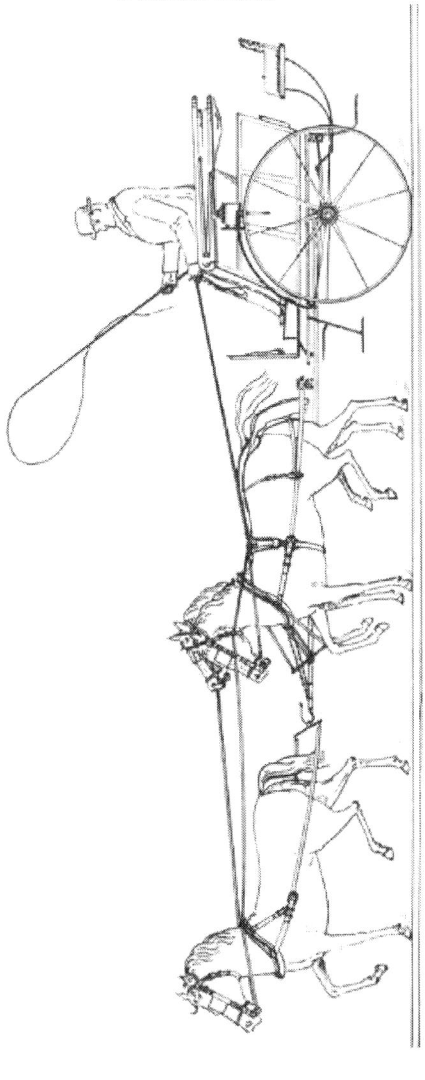

HOODED BUGGY, WITH PICK-AXE TEAM, CAPE-HARNESSED.

PLATE XXIII

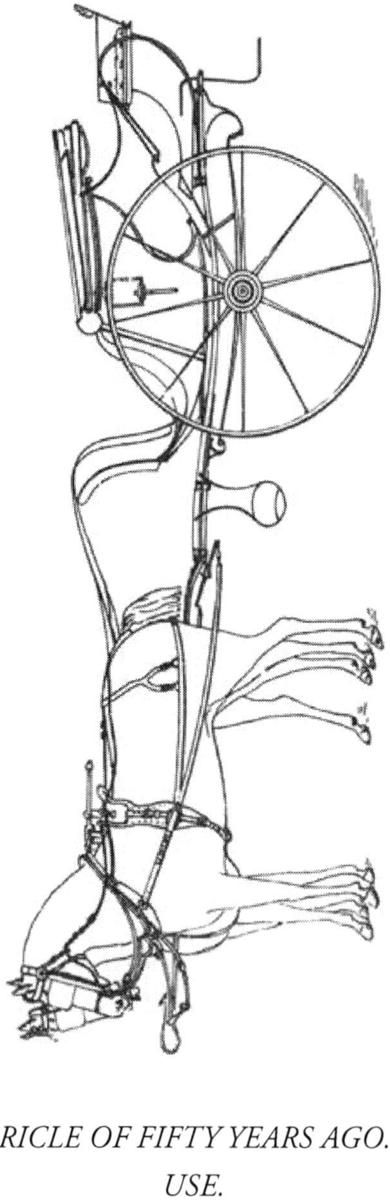

THE CURRICLE OF FIFTY YEARS AGO. STILL IN USE.

PLATE XXIV

ILLUSTRATIONS TO COLONEL PEDLER'S LETTER.

Perhaps, if he sees this, he may send you the average weight of such carts. I estimate this one at 8 cwt., but another competent judge tells me he thinks it is quite 11 cwt.

NIMSHIVICH.

Field, 22nd October, 1881:—

APE CARTS AND CURRICLES.

It is now some time ago that "Speed" put a question in the *Field* as to what superiority the Cape cart had over a curricle.

1.—The curricle is the most difficult tackle in use for putting the horses to, or for taking them out. The Cape cart is the easiest of any two-wheeled carriage, whether with one horse or two horses.

2.—It would be almost impossible to drive a team of four or more in a curricle, the wheelers would get sore backs and girth galls; in the Cape cart, it is simple, and a common practice, to drive four or six horses, and the only wrench on the wheelers, if any, is on the breeching.

3.—In case one horse falls, there is no strain on the harness of a Cape cart; if the same accident happens with a curricle, the bar must be pulled off the other horse's saddle. By the way, I have never been able to understand why the steel pin on a curricle saddle, on which the bar rests, is made so high, thus giving great leverage for breakage.

4.—If the pole straps are loose enough to allow of the horses going in the ruts (so very great a *desideratum* in rough countries), then, in case of a bad shy or any restiveness, one or both of the horses would inevitably get free from a curricle bar; and the consequence, at such a time, too, with horses already alarmed, may be imagined. With the Cape cart tackle, a horse cannot get away in this fashion; in short, the curricle harness is highly dangerous, and the other very safe.

5. It will often (indeed, always, unless the traces are of very accurate length) be found that the horses in a curricle are pulling by the saddle. It is common for them to keep the trap back, going down hill, by the saddle. These very great mistakes cannot happen with the Cape cart tackle. And there is no knee action.

I would strongly recommend breast straps for the Cape cart in preference to collars, the breeching acts so much better with breast straps. With a pole of very full length you may then go down any hill comfortably, although the pole straps are let out to enable the horses to go in the ruts. I sent "Nimshivich" a description of a harness of my own invention for a Cape cart, which harness is, I think, the lightest and simplest that could be used with safety. It would, of course, be possible to drive a Cape cart with nothing but collars and traces; but in case of the slightest kick a horse would get his leg over the traces, and, unless the pole was very long, the trap, going down steep hills, would run on to the horses if the pole straps were let out for ruts; but on a good road, and with the consequent short pole straps, the case would, of course, be no different from that of any ordinary four-wheeled carriage. I observe that the glorious fire brigade drive without trace lifters; but they have a very workmanlike-looking strap on the hames territt, which answers the purpose sufficiently, as their horses, though in the pink of condition, get the kick taken out of them in the first half-mile. The fire brigade tackle is, to my eye, the most workmanlike to be seen anywhere; and it would, under certain circumstances, be perfectly adapted to the Cape cart—the neck bugle being a light steel bar running through the pole strap ring on the hames, and having a pin at each end to prevent its slipping out of the ring. The pin would, of course, be removable, to take the horses out or put them to, and would be kept from falling out by a small leather strap and button.

GLENCAIRN.

P.S.—The seat of a Cape cart must be made to slide, so that a balance may be preserved at all times.

12th November, 1881:—

DRIVING CURRICLE.

"Glencairn" can perhaps tell me how to overcome three difficulties when driving curricle. I use a curved steel bar under the ponies, such as was in general use for our galloping mail gigs fifty years ago. This bar works in a holder, shaped like a tuning fork, the head of the holder passing through a hole in the pole. The bar has vertical and horizontal play: 1st, unless the ponies trot exactly in unison the cart joggles so as to nearly shake your teeth out; 2nd, when holding back, the bar strikes the ponies behind the elbows, making them back or threaten to lie down (I have no breeching); and 3rd, the backhands, though working easily through the saddles, have a tendency to jerk the pads in towards the pole. A friend showed me his Cape fittings—two half circles joined to a ring working on the pole, but it did not give him satisfaction. Any hints will oblige.

DUNEDIN.

The answer to "Dunedin" only appeared in the *Field*, or January 7th, 1882, owing to the clever writer "Glencairn" being absent from England:—

HEAVY TWO-HORSE CARTS.

The plan mentioned by "Dunedin" in the *Field* of November 12th is called a "belly bugle." It is not adapted for trotting work; it is used for heavy loads, dragged at a walk, because a very heavy load would require constant shifting with a neck bugle, unless the road was quite level.

It is many years since I was in South Africa, but I clearly recollect that the belly bugle was only used for heavy and slow work, for which purpose it is, in my opinion, much inferior to a plan—a sort of modified and very safe curricle—of which "Nimshivich" has drawings.

I here beg to say that there is no occasion to speak of Cape carts; any cart will do, if the seat is made to slide, and the wheels are of suitable height for the horses. The proper and explanatory term would be Cape "harness." I refer to light fast work and the neck bugle. The belly bugle is also a South African tackle, but there is no occasion for a sliding seat or any particular height of wheel with it.

GLENCAIRN.

In December, 1881, there was some further correspondence in the *Field*, between a "Conservative Whip" who advocated tandem as the best application of horse-power, and the author who maintained that the Cape system was better. This drew forth the following practical letter:—

TWO-WHEELED CART WITH PAIR OF HORSES.

I notice in your paper some difference of opinion as to the merits of driving a pair of horses poled to a two-wheeled cart. I have had a good deal of experience with that kind of team, and find it answers very well indeed, there being less and easier motion than with shafts; and I consider it safer. I tackle mine with yoke attached to pole in front of the horses. The yoke is hung with straps to top of the collars; the cart is kept from tilting by martingales which divide under the horses, and snap on each side of the breeching; thus allowing a good hold back when going downhill. The cart should be properly balanced with shifting seats. An ordinary light pole answers. Tackled thus, I can drive horses of different sizes, but the cart must not be very high. I use pole chains.—24th Dec., 1881.

A. V.

The lithographed designs for a Cape cart and harness accompanying these remarks have been made for me by Messrs. Atkinson and Philipson, whose experience and judgment have been of the greatest service to me in accurately representing the kind of vehicle that I think may be introduced with advantage both at home and in the colonies; see especially Plate XVII.

TURF HOTEL, NEWCASTLE.—FROM THE ORIGINAL WOOD BLOCK.

* Colonel J. P. Pedler.

APPENDIX.

DIRECTIONS TO COACHMEN—THEIR DUTIES—THEIR DRESS—HINTS ON DRIVING, &C.

IN 1796, one Felton, a coach-maker, made some observations, the common sense of which will be my best excuse for reproducing them here:—

It is by the coachman gentlemen are usually biased in what is to be done in the repairs or alterations of the carriage; and who from interested motives or capricious whims often go to extravagant lengths, abusing the implicit confidence their masters place in them, not only to the sacrifice of their property, but to the injury of the carriage, which often becomes a kind of property to the coachman or coachmaker, and the proprietor a dupe to one, or both, of their artifices. Coachmakers are too frequently made subservient to the coachmen, owing to the influence they have with their employers, and are therefore obliged to countenance the impropriety of their orders if they wish to preserve their customer. Therefore by stating what the probable expense for repairs may be for any length of time, a person may judge whether or not he has been taken advantage of by one or the other. It is in consequence of frequent and expensive repairs, which gentlemen often experience, that induces them rather to job or hire than purchase a carriage, supposing by that means to know the extent of their expenses; but in that they are frequently disappointed, by a number of charges for extras and what is termed accidents, which often exceed the expense of their own carriage.

I omit Felton's estimate, which would not be applicable now.

It is also essential to know in what manner a carriage may be best preserved, both in strength and beauty; for by neglect or ignorance a carriage is as much injured as by use. This depends entirely on the care of the servant, and it is a matter of consequence to the proprietor to examine as scrupulously into the merits of the coachman respecting his knowledge of a carriage, as it is of his experience among horses or his skill in driving them. A proper qualified coachman, made independent of his customary emoluments, is as valuable a servant as any in a gentleman's family; but otherwise, is as likely to be one of the worst, as his expectation of perquisites often produces extravagant expenses to his master and brings discredit on the builder; it is therefore, for a gentleman to consider his own advantage, and make an adequate allowance to his servant for the emoluments he deprives him of, whereby he may secure to himself treble advantages.

Felton observes on the subject of coachman's tools:—

It is very impolitic in many people not to furnish their servants with proper conveniences, for want of which they frequently incur treble the expense in one year as would supply, for many, all the necessary requisites for the coachman's use, with which he might employ many of those leisure hours while the carriage is unemployed, equally to the improvement of himself and the benefit of his master.

The requisites whereby a servant can make himself useful ought not to be denied him.

These are, I quote now from the latest information, sponges, chamois leathers, dusters,* water-brush, clothes brush, paint brush and bottle of black paint, pails, setter, oil, turpentine, rotten stone, whiting, powdered charcoal, washers' leggings, water hose of India-rubber tubing, 3 to 6 yards long with rose end where the cistern is so placed as to allow a pressure of water. Notice well that I have omitted an article which is found in almost every stable, viz. a spoke brush. Such an article ought never to be allowed to enter an establishment. Use one of these before every particle of mud and road grit is removed (how seldom is this the case) and you by degrees produce on the *soft* varnish the same dull surface that glass manufacturers obtain on their *hard* glass by a similar process. The following tools are also useful:—S. wrench, hammer, and wheel wrench. The latter should be found in every stable, but not for the purpose of a coachman oiling up his wheels. There are few men in the best coach making establishments who are allowed to oil up wheels till they are thoroughly acquainted with the work. The greatest care is required to so adjust the washers, nuts, &c., that there may be no undue wear on the various parts, which soon causes permanent rattling in the axle boxes. Felton observes that the coachmaker is usually applied to for the most trifling job which the servant might do if he had the materials. Certainly we cannot lay down a hard and fast line where the duties of the coachman end, and those of the coachbuilder begin. It must depend on the handiness and good-will of your servant. At any rate he should have no excuse in the lack of tools. On cleaning carriages Felton says:—

A carriage ought always to be cleaned, if possible, immediately after use, before the dirt dries on it; but when that cannot be effected, and the dirt is stiff, sluice it well and give the water time to loosen it, as by rubbing it when dry the small stones will scratch the paint and particularly injure the varnish on the panels. In placing the setter to clean and raise the wheels care should be taken that it does not press on or rub against any part, but that the bearing may rest on the pin only, as frequently the timbers are much bruised by this inattention. Soft water is preferable, if to be obtained; sea water is a great hurt to the paint, as the salt penetrates through the colour, and on the ironwork it totally destroys the paint and leaves the iron bare and rusty. The ironwork should always be wiped perfectly dry, particularly the spring, as the wet or damp gets between the plates, which occasions them to rust and to blister out at the edges.

The following "Hints for the Preservation of a Carriage" are from the *Hub*, of New York. They repeat, in some instances, what has been said elsewhere, but they are terse, neat, and pregnant, and would not bear reduction:—

A carriage should be kept in an airy, dry coach-house, with a moderate amount of light; otherwise the colours will be destroyed.

There should be no communication between the stables and the coach-house. The manure-heap or pit should also be kept as far away as possible. Ammonia cracks varnish, and fades the colours both of the painting and lining.

Whenever standing for days together, a carriage should always have on it a large cotton cover, sufficiently strong to keep of the dust without excluding the light. Dust, when allowed to settle on a carriage, eats into the varnish. Care should be taken to keep this cover *dry*.

When a carriage is new or newly-painted, it is better for it to stand a few weeks before being used. It will stain or spot, even then, unless care be taken to remove the mud before it dries on, or as soon afterwards as possible. A carriage should never, under any circumstances, be put away *dirty*.

In washing a carriage, keep it out of the sun, and have the lever of the "setts" covered with leather. *Use plenty of water*, taking great care that it is not driven into the body, to the injury of the lining. Use for the body a large, soft sponge; when saturated squeeze this over the panels, and by the flowing down of the water the dirt will soften and harmlessly run of; then finish with a soft chamois leather.

The same remarks apply to the under-parts and wheels. Never use a "spoke brush," which, in conjunction with the grit from the road, acts like sandpaper on the varnish, scratching it, and of course removing the gloss. If persisted in, it will rub of the varnish and paint even down to the wood. Never allow water to dry itself on a carriage as it invariably leaves stains.

Enamelled-leather heads and aprons should be washed with soap and water, and then rubbed very lightly with linseed oil.

In cleaning brass or silver, no acid, mercury, or grit should be used; the polish should be obtained solely by friction.

To prevent or destroy moths in woollen linings, use turpentine and camphor. In a close carriage the evaporation from this mixture, when placed in a saucer, the glasses being up, is a certain cure.

Be careful to grease the bearings of the fore-carriage so as to allow it to turn freely. If it turn with difficulty, the shafts or pole will be liable to strain or break.

Examine the carriage occasionally, and whenever a bolt or clip appears to be getting loose tighten it up with a wrench, and always have little repairs done *at once*. Should the tires of the wheels get at all slack, so that the joints of the felloes are seen, have them immediately contracted, or the wheels may be permanently injured. "A stitch in time saves nine!"

Examine the axles; keep them well oiled, and see that the washers are in good order. Castor oil is considered the best for oiling; sperm oil will answer; but *never use sweet oil*, as it will gum up. Be careful in taking off the axle-nuts not to cross the thread or strain them when replaced.

Keep a small bottle of black japan and a brush always handy, to paint the treads and steps when worn by the feet; nothing helps more than this to keep a carriage looking tidy. Lay on the japan as thin as possible.

Never draw a carriage out of, or back it into, a coach-house with the horses attached, as more accidents occur from this than any other cause.

Headed carriages should never stand with the head down; and aprons of every kind should be frequently unfolded or they will soon spoil.

As a general rule, a carriage, *with gentle work*, retains its freshness better than if standing for long periods in a coach-house. If the latter be necessary, draw the carriage out occasionally to air.

See that the coach-house doors can be so fastened as not to blow to by the wind.

Do not use oil cloths in carriages.—Covering the bottoms of carriages with oil cloths is a fruitful source of trouble, causing, as it does, the bottoms to rot by retaining the moisture between the wood and the cloth. Any article that is air-tight should not be used unless care is taken to remove it when the carriage is not in use. Grass mats are more cleanly and comfortable and less injurious.

THE GROOM OR COACHMAN'S DRESS.

Having guided the coachman in his duties let a word or two be added on his dress. A perfect equipage—meaning thereby, carriage, horses, harness and servants—necessitates that the latter should not be slovenly in their attire.

Seeking information on the subject of liveries, more particularly as appertaining to breeches, the department in which so many fail, the author applied to Messrs. E. Tautz and Sons, 495, Oxford Street, (not to be confounded with any other firm) and this is what they say:—

1.—We may make two pair of breeches for two men who take the same fit and one shall put them on properly and the other like a sloven.

2.—As a principle, we exact that all men who wear breeches shall be careful to observe the following rules:—*a*. Their breeches' buttons must run perfectly straight down the front of the leg as near the shin bone as possible. *b*.—In putting their breeches on they shall be careful to pull them up quite close under the knee so that the top buttons are placed as high as possible. *c*.—That four buttons be always allowed to show above the top, and that their boots be made sufficiently long to admit of this being done with a "breeches-knee" as short as possible. Nothing makes a man's leg look so short-legged as a long knee to breeches and a short boot.

3.—Captain Cooper of the Dorking Coach, one of our oldest customers, who was a great authority upon such matters, used to say that nature had provided man with an indication when the top buttons should go in the shape of the small hollow, or knee-pan, which is found immediately under the knee-cap, which not only tells us how high the breeches should be pulled up, but shows the *line (well in front)* which the four buttons above the boot should take. The accompanying sketches show this roughly, but we venture to say accurately.

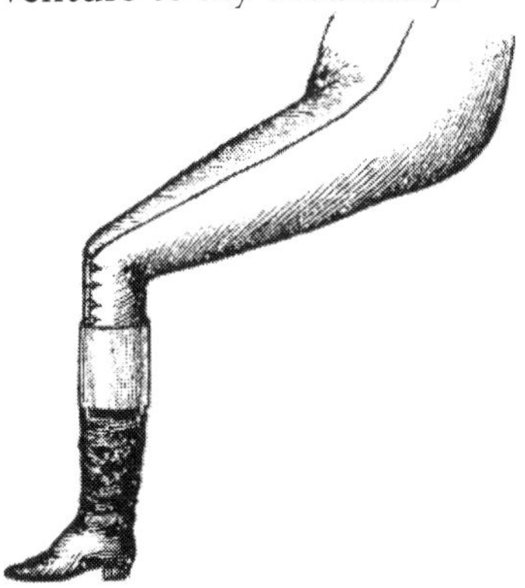

Smart.

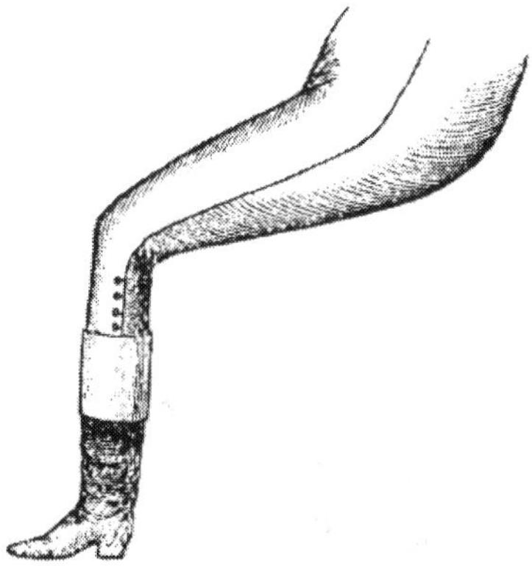

Slovenly.

4.—Following this, as a consequence, we may say that the shorter from the knee to the top of the top-boot the smarter a man's leg will look, and the breeches makers' art is quite wasted unless the wearer is careful to see that these rules are observed.

5.—Folding an overcoat so that it hangs neatly over the rail is very important. It should be buttoned first, then laid on the table in the servants' hall with the buttons next to the table, it should then be smoothed and the sleeves carefully laid at length and so turned over that no difficulty or hitch occurs in permitting the smooth breast part to hang down outside.

To what Messrs. Tautz and Sons have said about folding, the author will only add that, for some reason which he is at a loss to explain, the French coachman or groom has a knack of neatly folding his overcoat which our English servants would do well to copy.

HOW TO DRIVE.

As in the case of most English trades, the old-fashioned practice of apprenticeship is much more needed for the man who would become an efficient whip than reading written precepts. So I will refer the tyro to Mr. Houghton's remarks below, and put down here only those few practical hints which experience has taught the writer must ever be retained in the mind of those, whether of gentle or simple, who seek to drive horses properly.

1.—Driving gloves should be a full size larger than such as you would otherwise wear. Without this precaution you are liable to have your fingers most seriously cramped. Always have a pair of worsted gloves about you in case of rain, they are the only things through which the reins do not slip.

2.—If it is your object to learn to drive four-in-hand, do not attempt to mount the box until you have practised to make a double thong, which you will do best by mounting on a chair in the garden and placing four strings or reins in your left hand in their proper position, *i.e.* the fore finger between the leaders' reins, the near rein being above it; the second finger between the wheelers' reins, the near wheel rein being above it and next to and below the off leader's rein. You will then with the whip in your right hand practise making a double thong. This double thong, to give a rough and general indication, is the result of smartly striking the centre of the curve formed by the thong when the end of it is in your hand with the crop or end of the stick. That is one way. Another is by throwing the curve upwards and catching it with the stick. The same principle underlies both methods. It is the sharp collision of the curved thong and the end of the stick that produces the double thong. Unless you can make it easily and well you cannot drive four horses like a workman.

3.—Mr. Birch-Reynardson in his "Down the Road," says:—

> Always have an apron on your box; it hides a bad seat and a pair of bent knees. Sit straight on your box, with your elbows close to your side, your hands well down, your shoulders well back, your head erect, and your eyes well in front of you. Do not set your back up like a "pig in a rage," and do not bend over your footboard as if you were looking after the stump of the cigar you have just dropped. Above all, do not sit with your knees bent. Start quietly off the crown of a hill. If your pole-chain breaks, look out for loose gravel or fresh broken stones; they are a wonderful help in times of need. Always have a chain-trace in your boot.

To which may be added:—

> Never crack your whip—punish a lazy one—don't disturb those that are doing their work well. Also would I add, in the interests of the fair sex, that a coachman should point the whip, when in a state of rest, towards, but a little behind, the shoulder of the near wheeler, and not, as I have too often seen, in a line parallel with the box seat, whereby the parasol, and sometimes even the bonnet, of the fair passenger is disturbed. A shy woman will say nothing, but be rendered uncomfortable by such a display of ineptitude.

The following observations by the editor of the *Hub* (New York), form as complete a treatise on the science of driving as any writer has hitherto put together:—

SUGGESTIONS ABOUT HARNESSING.

First of all the horses should be measured for their collars, which should be made to fit exactly, or they will otherwise be sure to wring the horses. Some mark the collars as follows:—"n. 1." for near-lead, "o. 1." for off-lead, "n. w." for near-wheel, and "o. w." for off wheel. When a horse has worked for some time the collar takes the shape of his neck and becomes easier for him. A horse should have his own collar as much as a man should have his own boots, and in case he is made to shift from place to place—as some coachmen prefer that he should do—he should always carry his collar with him.

The rules for fitting the bits and bridles should be the same as in harnessing tandem. The lead-reins are often run through head terrets on the wheelers' bridles, being considered by some as good form for the park, but others consider this objectionable, as it throws weight on the wheelers' heads, and a hard-pulling leader under these circumstances must naturally worry the horse behind him. The simplest and best way, when you have a pulling leader, is to let the lead-rein pass underneath the throat-lash. Some may ask, "Why not pass them through a D or ring below the rosette?" to which we would reply that, passing underneath the throat-lash will save the wheel-horse from all worry and bother from his leader's hard pulling, or any other troublesome movement the latter may be guilty of.

The mode of attaching the lead-bars to the poles should be as in England, namely, with the crab at the end of the pole, and not with a hook underneath, as we have seen in some French coaches. The latter might do where there were three leaders abreast, as were formerly driven in France (frequently from a single bar); there is but one opinion on this subject among coaching men in England at the present time, and this is in favour of a crab at the pole end. Attention should be paid to having the draw-bar of the drag clear the bands on the front wheels, and the measurement from the outside of one bar to the outside of the other (of lead-bars) should fully equal that of the draw-bar, so that where your leaders go your drag will follow.

Reynardson advises the use of bearing reins, saying:—"Your team will look smarter, and you will have more comfort, for all four horses seldom carry their heads right, and it is a great nuisance for the driver to have to carry his horses' heads, to prevent them from lolling against the pole-hook, and perhaps catching their bits in the pole-chains, which I have seen occur more than once." We object to bearing-reins on principle, as giving the horse too little freedom of his head, and being liable to choke him in going up hill, as has happened to more than one horse. For road use the best authorities agree upon the absence of the bearing-rein except for park use.

In regard to buckled reins, Reynardson says:—"Formerly all reins were buckled, but gradually it became the fashion to let them hang down unbuckled, and then to have no buckles at all, and for twenty years I never saw such a thing as a buckle. It's a safe plan, nevertheless, to buckle your reins; and with the thin, dandy, new, slippery reins of the present day it is positively necessary, or they will easily slip through your fingers." A member of the New York Coaching Club suggests a convenient substitute for buckling reins, which commends itself to us. He says:—"As you have spoken of the buckles on the reins, I will say that for a long time, in all my driving, I have had my reins made with two loops, just like that which is always behind the buckle, sewed on to the rein in place of the buckle, one at the end where the buckle usually is, the other about five inches behind it. The end of the other rein passes through these loops, fitting them so tightly that a small amount of force is necessary to pull it out. This serves all the purposes of a buckle in case a rein is accidently dropped, and can be undone in an instant by a pull. In my opinion it solves the question of 'buckle or no buckle' entirely."

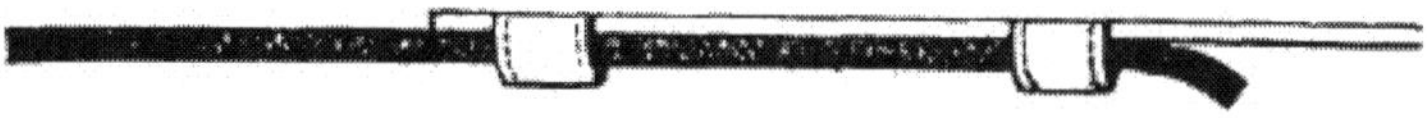

Have your reins long enough, strong enough, and broad enough—not less than a full inch in width; they will then be stiff enough for you to slip them up between your fingers, instead of pulling at them from behind your hand. Never omit to take a pair of worsted gloves with you, as, on a wet day, they are the only preventive of reins slipping.

HINTS TO FOUR-HORSE COACHMEN.

The following points have in part been suggested by Malet's "Annals of the Road," Reynardson's "Down the Road," and Sidney's "Elementary Hints," gathered by him from Tim Carter, the consulting counsel of the younger members of the London Driving Clubs; but we have introduced many additional hints gathered from some of the four-horse drivers in this country, including Jack Reford, Fownes, Jennings, George Irvine, and members of the New York "Coaching Club," which we hope will be found of value.

A four-horse coachman needs to have courage, decision, good eyesight, a flexible hand, and strength in the arms and back.

As a beginner, he should first practise with a coach and pair, and should wait until he has these under perfect command before the leaders are hooked on. The mere exertion of holding four horses going freely is so great that a pupil who has not developed the muscles of his left arm, by the use of Indian clubs or other gymnastic exercise, will very likely find himself disabled before he has finished his first lesson of fifteen minutes. We know this by experience. The preparatory training of the muscles of the left arm is, therefore, highly important, especially in the case of slight men, who, in this art, are at a disadvantage as compared with muscular or heavy men.

The gentleman who desires to become proficient in four-horse driving should first obtain the services of a teacher, choosing some really good coachman who has had long experience and fully understands the business; if he has learned his art in England, all the better. From him you will be able to pick up a multitude of points that will be of constant service to you, and indeed, without such instructions at the start, and from a competent teacher, you will probably always lack that confidence, knowledge of details, and elegance which every gentleman coachman ought to have if he would be distinguished from the common hack driver. Great care should be taken at the start to acquire proper habits, for nothing is so difficult to cure as a bad style.

The first few lessons will be well employed by sitting beside your teacher on the box, he holding the reins and you watching his every movement and learning by heart his instructions. The information thus gained will be of lifelong service to you as a coachman.

The next lessons should be devoted to practising how to mount, how to sit, how to hold the reins, how to start, how to turn to the right and left, how to stop, and how to dismount. We will endeavour, further on, to briefly explain each of these processes.

I.—MAKING UP YOUR TEAM AND HARNESSING.

The leaders should be of about the same size and height as the wheelers, though some prefer that the former should be taller. Sidney tells us that the most expensive four-horse teams in England match in height, action, character, colour, and age; and as to age, he adds, that "some of the best horses in London teams are perfect patriarchs," but then, they have never done any real work. The leaders should have courage and style and be both free and docile.

Temper in harness-horses is much to be regarded. Most coachmen agree that a horse should never know his place, but that he should go wheeler or leader, and on both sides; Nimrod, on the other hand, strongly advises that, "if a horse, working constantly in a coach, *fancies* his place, he should have it, and in nine cases out of ten he will pay for the indulgence."

Wheel-horses have the hardest place, as they are at work up hill or down; nevertheless, if favour be shown it must be to the leaders, for you may drag a tired wheeler home, but if a leader cuts it, you are planted. Always put your freest horse on the near side, as you will then have him better in hand than if he were on the off-side.

If a leader is weak, and cannot take his bar, pole up the wheeler that follows him, and the latter will place him by the side of his partner.

A horse with a tender mouth is difficult to handle in coach harness, particularly if he be a leader. A ring-snaffle is a safe bit for such a horse on all occasions.

In case of a hard-mouthed horse, try a well-twisted ring-snaffle—that will generally bring him to his bearings. This bit may not look well enough for use on a private drag, in which case the coachman should consult his own ingenuity in devising something to relieve the weight from his arms. We object to severe curbs, partly on the score of cruelty, and partly from having seen cases in *which a horse would pull against a curb, no matter how severe, and where, on substituting a light hit, no further trouble was experienced by the driver.* Make each horse as comfortable as possible; he will appreciate it, and show it by his work.

A hard puller is generally safest, and more in place at the wheel, and he will pull less with a free partner than with a slack one.

A kicking wheeler should be put on the near side, where he will be more under the control of the coachman. With gentlemen's teams, very fresh in condition, it often happens that a wheeler will kick over his trace, especially in a turn; this may be prevented by adding a light hip strap, which, when the horse rises, will take the trace up with him.

II.—MOUNTING.

Before getting upon the box, walk around your horses' heads, and see that the curb chains and coupling reins are all right, and, above all things, that the tongues of the billet buckles are secure. Many bad accidents have arisen from want of this precaution, and no man can be considered a scientific or even safe coachman who does not see personally to these things.

See also that the wheelers are not too closely poled up, and that their noses are not pulled too closely together by reason of the coupling reins being too short, as horses with their heads too close will neither go with comfort to themselves nor to him who works them.

The coachman should always mount from the off side.

Before mounting take your reins in the right hand, in the proper position, and, of all things, see that the ends are buckled or otherwise attached, as dropping a rein has caused more than one bad accident. If the whip is not in the socket, it should also be taken in the right hand with the reins. This leaves the left hand free to aid you in mounting, which should be done as follows:—Left foot on the hub, right to roller-bolt, then left to step on front boot, thus bringing your right foot on to the foot-board, which lands you on the box.

Having fully satisfied yourself that every horse is properly harnessed and bitted, mount the box deliberately, your groom or grooms being at the horses' heads meantime, and mind that the team stands still until you give the word to start. This is an essential part of the breaking of a gentleman's team that is too often neglected, for it is essential for the comfort of the driver that a team, however high-couraged, should wait until they hear his word, "All right, my beauties!" or, "Now, Kittens!" or other similar order, before they start.

Having reached the box, change the reins from the right to the left hand, in proper position, seat yourself comfortably, and see that all is right. Do not take your whip out of the socket until your reins are adjusted.

III.—HOW TO SIT.

A graceful, and at the same time a firm, seat on the box is a great set-off to a coachman. He should sit as easy as possible, with his legs well before him, but not too stiff as if doing hard work, for they would then become cramped and uncomfortable; and with the exception of a pliant motion of his loins, on any jolting of the coach, his body should be quite at rest, even when he applies the silk. In handling the ribbons, as little motion of the arm should be made as possible.

The driver should have a high seat, overlooking and commanding his horses, but it is a mistake, and one that young coachmen are particularly guilty of, to sit too erect and stiff. It is essential that, in the event of a horse falling or the team attempting to bolt, the driver should be able to instantly use his utmost strength and weight to hold them and at the same time put on the brake, and this he cannot do if he is standing bolt upright against a sloping cushion, according to the mistaken fashion of to-day.

IV.—HOW TO HOLD THE REINS.

The mode of holding the reins and guiding the horses is practically the same as with a tandem team. Place the reins as follows:—Slip your forefinger between the lead-reins, near side over the finger; and second finger between the wheel-reins, near side on top.

Never divide your reins under any circumstances, as it is the worst of bad form, and uncoachmanlike.

Having taken your seat, adjusted the apron and your reins, next take care to have the leaders so in hand that, when they start off, they will be clear of the splinter bars.

V.—HOW TO START.

Let your wheelers in every case start the coach, and the leaders will come to their work as soon as they feel that the wheelers are under way.

On starting, the reins must be placed, and throughout the drive retained, at such a length and in such position that you can pull up your team at any moment.

The wheelers should start the coach, and turn it round, without the leaders ever feeling their traces, and whenever the wheelers stop it, the traces of the leaders should be slack.

The next important matter to be attended to is to get the four horses to start together and work together. To effect this, avoid the use of the whip at the start, as this tends to excite the horses; start them with a chirrup; when once on the move gather them by equalizing your reins and get the horses straight. The only case in which a whip is ever necessary in starting, is where one or more of your horses will not face the bit, which often causes trouble at the start, and this should be cured by words of encouragement rather than by blows.

Avoid this one common cause of trouble with horses at the start, namely, having the reins so short as to catch the leader's mouth before he can straighten and feel his traces, but rather let the reins be a little long until after the leaders are started, and then take them in hand; this rule applies to all four horses, but with particular force to the leaders, inasmuch as if they do not start off freely, the wheelers are unable to.

"He who makes a good beginning has half finished his work," and this applies to driving coach horses. Harness them well, and they are much easier driven. It is also with coach horses as with mankind: the physical strength is with the governed, and we must, therefore, humour them a little. When starting, don't pull at their heads, but feel their mouths lightly, or they may bolt, throw themselves down, or break through their harness. If old horses, and the stage commences with a hill to go down, let them feel their legs for a few hundred yards before they are put to their usual pace.

VI.—HOW TO TURN TO THE RIGHT AND LEFT.

In turning or inclining four horses, more pressure on the reins is of course necessary than with two horses.

When inclining the horses suddenly, as to give the road to a passing team, it is important that the wheelers and leaders be inclined at the same instant, in which case the wheelers and leaders will for a moment take up two different lines of direction.

In all other cases, be careful to make your wheelers go over the same ground as the leaders; observe the latter rule in turning at right angles, as into a cross-street or into a gateway.

When in a turn, point your leaders well, that is, take proper ground for them to make the turn, and let your wheelers follow them over the same ground. As wheel-horses are always in a hurry to make the turn, first "shoot" or steer them to the contrary side, just as you have first pointed your leaders; for instance, if your turn is to the right, catch up your near-wheel rein, or *vice versa*. This will keep the head of your pole—which you should have your eye upon—just between your leaders, and your wheelers will follow as if they were running on a straight road. This also aids you to clear your coach of posts, gutters, &c. No man can make a neat turn with four horses unless he "shoots" his wheelers at the same time that he points his leaders.

Never turn a loaded coach short, even at a slow pace, for it is never safe when there is not an even bearing on the transom beds. Brewster & Co., of New York, have introduced a simple improvement which, in a great measure, obviates the danger named; but even with this aid you cannot be too careful.

Although there are certain rules for passing and meeting other carriages on the road, yet there are times when they need not be strictly adhered to, and a little mutual civility and accommodation between coachmen is very desirable. Thus, if you have a hill in your favour, and a loaded coach is coming up at the same time, you ought, if you can do it with safety, to give the firmest side of the road to the other coach.

VII.—HOW TO STOP.

To pull up your team at a moment's notice is no easy matter to learn, and one lesson may properly be devoted to this. Practice, and practice alone, can teach you this; and unless you can do this, you have not proper control of your horses. Tim Carter says:—"If I get alongside of a young gentleman who has great conceit of his own driving, the first thing I ask him to do, is to stop, and that suddenly, when he least expects such a request."

The only rules that can be given about stopping are to put on your brake at once as hard as possible, and put all the strength of your arms to the wheelers, who will naturally pull up the leaders and stop the coach. Some coachmen use the voice, saying "Whoa!" at the same time, but this is not considered good form, unless your horses become unsteady; then it is well to speak.

VIII.—HOW TO DISMOUNT.

First put your whip in the socket. Retain the reins in the left hand, and let your right hand take a strong hold of the rail, then swing yourself round, putting your right foot on step on front boot, left foot on roller-bolt, then right hand to step, and lastly right foot on hub of wheel, which will bring you to the ground with your back to the coach, and your left hand well to your side. If this ends your journey, unbuckle your reins, and throw the near-side reins to the horse-keeper, who is supposed to be at the near-side ready to take the team in hand, retaining the off-side reins until the wheelers are uncoupled, and the leaders taken from the bars.

IX.—USE OF THE WHIP.

Having learned, be careful not to overdo the thing by making a display of your skill when there is no necessity for it. *The whip should be used as little as possible.*

Nimrod, in speaking of whipping coach-horses, says:—"One of the best proofs of a good coachman is to see his right arm still; and although, for the safety of his coach, he ought to be able to punish a horse when he requires punishment, yet he should, on all accounts, be as sparing of the whip as he can. Horses may be whipped till they become callous to whipping, and therefore slow; and in the condition in which coach-horses are now kept, a pound of Nottingham whip-cord will last a good coachman his lifetime. The very act of throwing the point of the thong over the wheelers' heads or letting it fall on their backs, as a fisherman throws his fly upon the stream, will set half the coach horses, in these days, into a gallop.

Remember, also, that a free leader should not be hit while making a short turn, or he may break his bar or the pole hook. Neither should leaders be hit when going over a bridge which is much raised, or when the pole points upward, as their draught on the end of it would be liable to snap it in the futchels.

Speaking to coach-horses from the box is nowadays considered bad form, but it is not without its effect. Nimrod gives an example of a terribly slow team which were cured by a new driver, who pounded them well with a broom handle before harnessing them for the trial trip, at the same time speaking to them in a certain rough language, and ever afterward it was only necessary for them to hear that voice to make their best paces.

X.—ACCIDENTS.

Birch Reynardson gives some excellent points or "dodges" under this head, which merit brief mention. He begins by saying:—"Never get into a mess if you can possibly keep out of one; 'discretion is the better part of valour;' there is no great art required to get into a mess, but there is sometimes much difficulty in keeping out of one." "A coach is a dangerous thing if all don't go right; and though I have seen men do their best to upset a coach without succeeding, still, when least expected, it's as easily done as upsetting an old woman's apple-stall."

Always carry a spare trace, or, what is better, a chain trace; it takes but little room in one of the boots, and often proves a friend. True, it is seldom wanted, but a trace sometimes goes when least expected, and in such a case, if you have not a spare trace of some kind, you're "in a fix," and if at night, "in a regular fix"—as helpless as a pump without a handle. The chain trace should be about a foot and a half longer than your ordinary leather trace, with a ring at one end, big enough for your chain to drop through, to make a loop to go over your roller-bolt, and a hook at the other end that will go into the links of your chain, that you may take your trace up or let it out, as you would do with a leather trace and a buckle.

The brake-block should be of hard wood (oak is preferable), twelve to fourteen inches deep, and with a curved bearing-surface. The smaller iron blocks, sometimes used, are not to be recommended, as they touch the wheel at only one point, and do not bite as wood does; their unserviceableness is evident from the fact that, in going down an ordinary hill, the skid has to be used as well; moreover the noise which iron blocks cause is very disagreeable—even painful to many. Be careful that your connecting bar be made stout, and in proportion to the size of your brake-block.

In case your horses run away with you, and you lose control of them, keep cool, steer them in the middle of the road, keep the wheelers away from the splinter-bar, and look out for a piece of gravel or loose stone road, to assist you in retarding them.

Recollect that if your pole chains go, or get loose, going down a hill, and no skid on, it is a somewhat serious business.

XI.—DRIVING A PICKAXE TEAM.

Birch Reynardson says:—"Should you come to grief with one of your horses—should one fall down dead, or get the staggers, or break one of his legs, or from some cause be so full of pain that you are obliged to take him out of the coach, and drive home without him, of course you must leave him on the road, and make a 'unicorn,' or what we used to call a 'pickaxe team.' For making the said 'pickaxe,' particularly if going a long journey, it is a prudent measure to carry a spare small bar, with an eye or ring set contrary way to the usual eye on a small bar. The ordinary eye will not let your small bar sit the right way on your pole-hook. As a set of spare bars is generally carried behind a drag, a bar of the description I name may as well be carried instead of one of the ordinary small bars, as you never require more than one of either kind, and the bar I have named is the only useful one in case of the sort of accident I have mentioned. Should you put your main bar on, you would find that, from the span being wider in the main bar than in the ordinary bar, your traces would not be long enough, and if you did get them on, when your leader was pulling hard up hill, the hooks of your main bar would be liable to be broken. This I have seen happen, and though, luckily, one is not often called upon to drive a 'pickaxe,' the wrinkle is worth knowing. In coaching days of old, a bar such as I have described was generally to be found in the hind boot.

Having learned the above points, day after day should be devoted to their practice, for successive weeks, until the pupil performs the requisite movements instantly and mechanically. It would be well for you to practice at first with an old team that have learned to obey the slightest indication and to do half the driver's work for him, but a man is not a *coachman* until he can hold, turn and stop, fresh, fiery horses, not all of the same temperament, and feel perfect confidence of his mastership over them.

Practice first on level roads, clear of traffic; then try crowded thoroughfares and steep hilly roads, where the duty devolving upon you will be materially increased and new occasions for "points" from your instructor will constantly present themselves. The pupil should never be tired of asking questions, and the instructor never tired of answering them.

NEWCASTLE-UPON-TYNE: ANDREW REID, PRINTING COURT BUILDINGS, AKENSIDE HILL

* Gard's plate cloths are worth attention here, because, after being used for cleaning and polishing bright metal parts, they serve as dusters afterwards.

26357460R00123

Printed in Great Britain
by Amazon